Engineering Failure Analysis

Edited by Kary Thanapalan

Published in London, United Kingdom

IntechOpen

Supporting open minds since 2005

Engineering Failure Analysis
http://dx.doi.org/10.5772/intechopen.84666
Edited by Kary Thanapalan

Contributors
Eskinder Desta, Gashaw Assefa, Guodong Cui, Shuaijiang Yan, Chengsong Zhang, Chuan Yang, Dazhi Chen, Juan Rodrigo Laguna Camacho, Silvia Marina Sánchez-Yañez, Maria Ines Cruz-Orduña, Gabriel Juárez-Morales, Luz Maria Ramos-Gonzalez, Cristóbal Cortez-Domínguez, Roberto Cabrera-Santiago, Javier Calderón-Sánchez, Alireza Khalifeh, Anantha Padmanaban D, George Kordas, Lei-Yong Jiang, Prakash Patnaik, Alexey G. Kolmakov, Michail L. Kheifetz, Roselita Fragoudakis, Vicente Romero, Amalia Black, George Orient, Bonnie Antoun

First published in London, United Kingdom, 2020 by IntechOpen
IntechOpen is the global imprint of INTECHOPEN LIMITED, registered in England and Wales, registration number: 11086078, 7th floor, 10 Lower Thames Street, London, EC3R 6AF, United Kingdom
Printed in Croatia

British Library Cataloguing-in-Publication Data
A catalogue record for this book is available from the British Library

Additional hard and PDF copies can be obtained from orders@intechopen.com

Engineering Failure Analysis
Edited by Kary Thanapalan
p. cm.
Print ISBN 978-1-78985-945-4
Online ISBN 978-1-78985-946-1
eBook (PDF) ISBN 978-1-78923-829-7

For EU product safety concerns:
IN TECH d.o.o., Prolaz Marije Krucifikse Kozulić 3, 51000 Rijeka, Croatia, info@intechopen.com or visit our website at intechopen.com.

Meet the editor

Dr Kary Thanapalan, Senior Lecturer & Head of Research in Mechanical Engineering, University of South Wales, UK. Dr Thanapalan received a B.Eng. (Hons) degree in control engineering from the City University London. Later he received a Ph.D. in aerospace control from the University of Liverpool, and then conducted his R & D activities in various capacities. He is currently working as a Senior Lecturer and a Head of Research in Mechanical Engineering at the University of South Wales. He is a chartered engineer and a member of IET (CEng, MIET). He is also a Senior Fellow of the Higher Education Academy (SFHEA). He is the Editor-in-Chief of the *Journal on Instrumentation and Control Engineering*. He serves as a member of TPC for many international conferences. His current research area includes systems engineering, control & automation, and material science.

Contents

Preface

Engineering material science is a diverse field that can be presented in many different ways. The title of this book, *Engineering failure analysis*, reflects the attitude of the present treatment. The book is about theory and practical applications, but the main focus is on theory with direct consequences for the understanding and practical use of available techniques. This book gives the reader an insight on basic principles to allow them to approach a practical problem.

The core material of the book should be suitable for a graduate-level course in engineering materials, in particular engineering failure analysis. As a prerequisite for such a course, it is expected, although not absolutely necessary, to require that the student should be familiar with basic mechanical properties of materials.

The structure of this book is as follows; initially, basic theories and definitions to failure modes, analysis and case studies are presented. This is followed by the topics addressed by the key chapters including: information on stress analysis – strengths and limitations of traditional theoretical approaches to FRP laminate design against failure; stress corrosion cracking behaviour of materials; failure analysis and durability issues. For example, one chapter describes fracture behaviour of solid-state welded joints in detail. The chapter presents a detailed analysis of practical and theoretical studies, also a methodology for characterising fracture behaviour of solid-state welded joints. Another chapter, *An effective approach to investigate turbine hot component failure,* investigates the influence of key factors such as variable fluid dynamics, turbine effect, nozzle guide vane, air cooling, and failure in such a system. The book contains a wide range of key topics within the field of engineering materials.

Finally, editing a book takes time, and I probably would not have been able to finish this one on time had I not had the privilege of allocated research activities time and I thank Ewen Constant, Head of Department of Aeronautical and Mechanical Engineering and Dr. Paul Davies, Head of School of Engineering for making this possible and providing inspiring working conditions. Support from Mateo Pulko, Author Service Manager, IntechOpen for coordinating and completing the book has also been important.

It was unavoidable that a lot of the editing of the book had to be done on overtime. I thank my family for letting me use their time.

Dr. Kary Thanapalan
University of South Wales,
UK

Section 1

Stress Analysis

Chapter 1

Strengths and Limitations of Traditional Theoretical Approaches to FRP Laminate Design against Failure

Roselita Fragoudakis

Abstract

The strength of Fiber Reinforced Plastic laminated structures is strongly dependent on the stacking sequence of the laminate, and consequently the fiber orientations of the individual laminae (also referred to as layers or plies). Classical Lamination Theory (CLT) is a theoretical tool providing the strain and stress distribution in a laminate based on its stacking sequence and material properties. On the other hand, first ply, and consequent ply failure can be approximated with interactive failure criteria, such as the Tsai-Hill and Tsai-Wu. Technological advances often require material alternatives to metallic structures, and FRPs constitute optimum solutions to such selections. However, these structures are no longer just plain laminates with unidirectional fibers in their laminae, they include geometric discontinuities allowing ease of assembly. Such discontinuities become stress concentration regions, which require extra attention upon design against failure. This chapter discusses the extent to which the traditional analysis of FRP failure, using CLT and interactive failure criteria is adequate in structures with discontinuities, and suggests extra analysis steps to be considered when designing against failure in the area of the discontinuity.

Keywords: fiber reinforced plastics, classical lamination theory (CLT), interactive failure criteria, linear fracture mechanics, stacking sequence, fiber orientations, fist ply failure, unidirectional fibers

1. Introduction

Industries are constantly turning towards new material alternatives that can provide lighter structures of high strength and customizable stiffness to the needs of the destined application. A polymeric matrix with an appropriate reinforcement comprises composite material solutions for a wide range of industries from the aeronautics and automotive to the battery industry.

A special case of such composite materials is Fiber Reinforced Plastics (FRPs). These composite materials have an epoxy resin matrix and a fibrous high-strength reinforcing phase. As a result, they provide high strength and stiffness, while being much lighter than any metal. Additionally, FRPs are highly corrosion resistant [1]. In a majority of applications FRP layers are laminated into beam like structures.

Therefore, they offer the option of being tailored to the desired properties of the destined application. FRP laminates of unidirectional laminae (i.e. long fiber reinforcement) can be optimized to have stacking sequences that will provide an optimum strength and stiffness at low weight.

The heterogeneous and highly anisotropic nature of composites, and consequently FRPs, is due to the fact that composite materials are made of two or more constituents that are insoluble in each other. As a result, the anisotropic nature of FRPs should always be considered when evaluating or predicting their failure mechanisms.

Failure in composite materials is defined as the point when the component seizes to perform adequately for the application it is designed for. At that point, failure may be described as catastrophic or simply degradation of the material properties. Understanding the mechanisms that lead to any type of undesired failure is very important when designing a component against failure.

In laminated FRPs there are ways to predict when failure will first occur in the laminate. First ply failure, will not always mean catastrophic or not failure of the composite, however, it will denote when failure is first observed in the laminate, and at which specific ply. It is possible that the FRP laminate will still function properly, as the load will be carried by the remaining plies. As a result, the design and choice of stacking sequence specify the maximum acceptable load for an application, and in the case of cyclic loading applications, can even specify which maximum applied load will cause first ply failure [2–5].

This chapter is divided into two sections; the first section discusses the heterogeneous and anisotropic nature of composites, and how Classical Lamination Theory (CLT) is used to determine the state of stress in unidirectional FRP laminates. Furthermore, two interactive failure criteria, the Tsai-Hill and Tsai-Wu, are discussed as the criteria of predicting first ply failure in FRP laminates. The above failure criteria are useful in conjunction with experiments in determining first ply failure in unidirectional FRP laminates. In the case a laminate contains a geometric discontinuity, such as a hole or tapered edge, the unidirectionality of the fibers is interrupted at the region of the discontinuity. As a result, the above failure criteria seize to accurately predict failure at the discontinuity, which additionally becomes a stress concentration region. The second section of this chapter, discusses the inadequacy of the above methods in designing against failure in laminated FRP components with such geometric discontinuities, and suggests additional analysis combining the Tsai-Wu failure criterion with fracture mechanics to better evaluate and predict failure in such regions.

2. Predicting failure in unidirectional laminated FRPs

2.1 Laminate stress distribution and classical lamination theory

The combination of the matrix and reinforcing phases (i.e. fibers in FRPs), which remain insoluble in each other, offer the composite material its anisotropic and heterogeneous nature, while at the same time a combination of the properties of both constituents. The volume percent of the total material occupied by the individual constituents (matrix and fibers) regulate the properties of the composite. As a result, the properties of FRPs may be tailored to the needs of an application by selecting the appropriate volume percent of fibers and matrix.

The anisotropic nature of the FRPs is mainly due to their reinforcing phase, the fibers, as the matrix phase is assumed to be homogenous and isotropic. The reinforcing phase may take the form of long unidirectional fibers, woven fibers, short, or chopped fibers that are scattered in the matrix. The theories and criteria considered in this chapter only concern long unidirectional fibers.

Laminated FRPs are structures composed of two or more FRP layers. These layers are also referred to as plies or laminae. Each lamina has its reinforcing phase oriented in a specified way and the fibers occupy a given volumetric fraction of the lamina. As a result, the properties of the laminate are determined by each lamina. The laminae are stacked together to create the laminate. The order of stacking is very important as the different orientations of the fibers in the individual laminae, as well as their volume percentage, affect the mechanical properties of the whole laminate. The selection of the order of laminae stacking is called the stacking sequence of the laminate.

Although it is often the case to view a FRP laminate as a homogeneous structure of isotropic bulk mechanical properties, this approach should only be followed for macroscopic analysis, when geometry and loading conditions are investigated rather than the specific effect of the material properties. When investigating the strength, stiffness, and designing against failure, the anisotropic and heterogeneous nature of the FRP laminate should be considered. In such cases, the analysis is in the lamina level or even a microscopic level of the individual constituents of the composite: the matrix, reinforcement, and their interface. The discussion that follows concerns the lamina level.

The FRP mechanical properties, although affected by the mechanical properties of its constituents, differ greatly from them and depend additionally on the volume fraction that each phase occupies. Rules of mixtures is the set of equations that calculate the elastic properties of a composite material taking into account the individual properties of its constituents and their volume fractions. The Young's moduli (E_i), shear moduli (G_{ij}), and Poisson's ratios (v_{ij}) are determined using Rules of Mixtures and the Halpin-Tsai equations. There exist therefore, three Young's moduli, one for each material direction (Eqs. (1) and (2)) and four shear moduli (Eq. (3)). To calculate Poisson's ratios the bulk moduli (K) are used (Eqs. (5)–(9)) [6].

$$E_1 = (1 - f)E_m + fE_f \tag{1}$$

$$E_2 = E_3 = E_m \frac{(1 + \xi\eta f)}{(1 - \eta f)} \tag{2}$$

$$G_{12} = G_{21} = G_{13} = G_{31} = \frac{G_m(1 + \xi\eta f)}{(1 - \eta f)} \tag{3}$$

where the subscripts m and f, refer to matrix and fiber, respectively, and 1,2,3, to the directionality of the material. The constant f is the volume fraction of fibers in the composite such that $0 \leq f \leq 1$, $\xi \approx 1$, and

$$\eta = \frac{\left(\frac{E}{E_m} - 1\right)}{\left(\frac{E_f}{E_m} + \xi\right)} \text{ or } \eta = \frac{\left(\frac{G}{G_m} - 1\right)}{\left(\frac{G_f}{G_m} + \xi\right)} \tag{4}$$

$$K = \left[\frac{f}{K_f} + \frac{(1 - f)}{\mathrm{K}_m}\right]^{-1} \tag{5}$$

$$K_f = \frac{E_f}{3(1 - 2\nu_f)} \tag{6}$$

$$K_m = \frac{E_m}{3(1 - 2\nu_m)} \tag{7}$$

$$\nu_{12} = (1 - f)\nu_m + f\nu_f \tag{8}$$

$$\nu_{23} = 1 - \nu_{21} - \frac{E_2}{3K} \tag{9}$$

To design against failure the FRP laminae should be built with an optimizable strength, and a desirable stiffness, while maintaining a low weight. The strength of the FRP laminate depends on that of the individual laminae, and can be optimized by choosing an appropriate stacking sequence. Therefore, although the above equations play an important role in determining the mechanical properties in different directions, it is important to start accounting for the orientation of the fibers in each lamina.

The constitutive relationships for FRPs use the generalized Hooke's Law (Eq. (10)). A total of 81 elastic constant would be required to fully characterize the FRP behavior. However, by assuming symmetric stresses and strains, the required elastic constants become 36. The lamina level contains two sets of axes that express the material direction; a set of local and a set of global axes. The local axes, also referred to as principal axes, have the longitudinal axis parallel to the longitudinal fibers. The global axes, are a reference frame of the laminate, where the horizontal, transverse, and normal directions coincide with the dimensional directions of the laminate. As a result, the longitudinal axis of the local reference system makes an angle with the global horizontal direction, thus allowing measuring the angle of the fiber orientation in each lamina. Consequently, each lamina has three mutually orthogonal axes of rotational symmetry, which further reduce the 36 elastic constant to 12. Only 9 of these constant are independent.

$$\sigma_{ij} = E_{ijkl}\varepsilon_{kl} \tag{10}$$

As seen from Eq. (2) of the Rules of Mixtures, the properties of the lamina in directions 2 and 3, the directions normal to the longitudinal fibers, are the same. As a result, the plane 23 of the lamina is an isotropy plane. Therefore, the FRP lamina characterized as transversely isotropic, a special case of orthotropic materials, requires just 5 independent elastic constants to fully determine its behavior. Classical Lamination Theory (CLT) used with orthotropic continuous laminated composite materials builds a set of equations that lead to the development of constitutive relationships that determine the state of stress in each layer [7–9]. CLT accounts for both the lamina orientations and its position in the laminate, showing therefore, the significance of the stacking sequence to the strength and performance of the laminate. To determine the position of a lamina in the laminate a common starting reference point of lamina numbering is the bottom layer. This bottom lamina, becomes lamina 1. There also exists a fictitious plane dividing the laminate in two equal half portions, called the mid-surface plane. This plane serves as a position datum for the laminae (**Figure 1**).

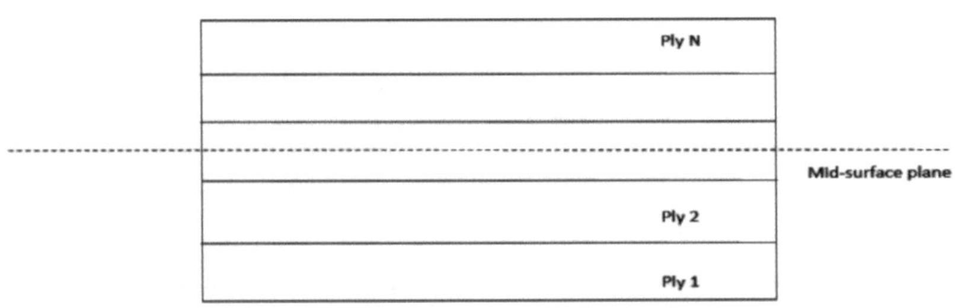

Figure 1.
Stacking sequence and nomenclature.

CLT builds constitutive relationships using elastic properties, which can be determined by Rules of Mixtures and the Halpin-Tsai equations or experimental data, as well as thermal expansion properties at each material direction. Depending on the nature of the fibers and the destined application of the structure, hygroscopic coefficients may also be considered. In the case of a transversely isotropic material, only two sets of material properties are required: one set in direction 1 and one set in either direction 2 or 3. CLT first evaluates a stiffness matrix (Q and $\overline{Q}$) for each lamina accounting for the elastic properties in the required directions and the orientation of the fibers. The overbar above Q denotes all off-axis laminae, i.e. those whose fibers make an angle with the global horizontal laminate direction, while the absence of a bar above Q refers to the stiffness matrix of on-axis laminae, i.e. those whose fibers are parallel to the global horizontal direction, having a 0° orientation. To distinguish between the different lamina stiffness matrices a subscript (k) is used, denoting the kth lamina in the laminate.

$$\overline{Q}_k = \begin{bmatrix} \overline{Q}_{11} & \overline{Q}_{12} & \overline{Q}_{16} \\ \overline{Q}_{21} & \overline{Q}_{22} & \overline{Q}_{26} \\ Q_{61} & Q_{62} & Q_{66} \end{bmatrix}$$

where

$$\overline{Q}_{11} = Q_{11}\cos^4\theta + 2(Q_{12} + 2Q_{66})\sin^2\theta\cos^2\theta + Q_{22}\sin^4\theta$$
$$\overline{Q}_{12} = \overline{Q}_{21} = (Q_{11} + Q_{22} - 4Q_{66})\sin^2\theta\cos^2\theta + Q_{22}(\sin^4\theta + \sin^4\theta)$$
$$\overline{Q}_{22} = Q_{11}\sin^4\theta + 2(Q_{12} + 2Q_{66})\sin^2\theta\cos^2\theta + Q_{22}\cos^4\theta$$
$$\overline{Q}_{16} = \overline{Q}_{61} = (Q_{11} - Q_{12} - 2Q_{66})sin\theta\cos^3\theta + (Q_{12} - Q_{22} + 2Q_{66})cos\theta\sin^3\theta$$
$$\overline{Q}_{26} = \overline{Q}_{62} = (Q_{11} - Q_{12} - 2Q_{66})cos\theta\sin^3\theta + (Q_{12} - Q_{22} + 2Q_{66})sin\theta\cos^3\theta$$
$$\overline{Q}_{66} = (Q_{11} + Q_{22} - 2Q_{12} - 2Q_{66})\sin^2\theta\cos^2\theta + Q_{66}(\sin^4\theta + \cos^4\theta)$$

where

$$m = \cos\theta. \text{ and } n = \sin\theta \tag{11}$$

$$[Q]_k = \begin{bmatrix} Q_{11} & Q_{12} & 0 \\ Q_{21} & Q_{22} & 0 \\ 0 & 0 & Q_{66} \end{bmatrix}$$

where

$$Q_{11} = \frac{E_1}{1 - \nu_{12}\nu_{21}}$$
$$Q_{12} = Q_{21} = \frac{\nu_{12}E_2}{1 - \nu_{12}\nu_{21}}$$
$$Q_{22} = \frac{E_2}{1 - \nu_{12}\nu_{21}}$$
$$Q_{66} = G_{12}$$

As mentioned above, CLT focuses on each lamina individually. The constitutive equation of the kth lamina (Eq. (12)) relates the stress distribution in the lamina to the lamina strain through the stiffness matrix. The strain distributions is a function of the mid-surface strains ($\varepsilon_{ij}{}^{o}$) and curvatures (κ_{ij}), which are common to all laminae of the laminate and depend on loading conditions. The effect of thermal (α_{ij}) and hygral effects (β_{ij}) is also included in the strain calculation as they are responsible for residual strains in the laminate that may be induced during the manufacturing and curing process or service life of the composite.

$$\begin{Bmatrix} \sigma_x \\ \sigma_y \\ \tau_{xy} \end{Bmatrix}_k = \left[\overline{Q}_k\right] \left(\begin{Bmatrix} \varepsilon_x^o \\ \varepsilon_y^o \\ \gamma_{xy}^o \end{Bmatrix} + z \begin{Bmatrix} \kappa_x \\ \kappa_y \\ \kappa_{xy} \end{Bmatrix} - \begin{Bmatrix} \alpha_x \\ \alpha_y \\ \alpha_{xy} \end{Bmatrix}_k \Delta T - \begin{Bmatrix} \beta_x \\ \beta_y \\ \beta_{xy} \end{Bmatrix}_k \overline{M} \right) \tag{12}$$

The above relationship is the stress–strain relationship of the kth lamina. In order to build relationships for the stress and strain distributions in the laminate, which can then be used to determine first ply failure and the strength of the whole laminate, the loading conditions of the laminate should be considered. Three matrices in CLT: the Extension Stiffness Matrix, (A_{ij}), the Extension-Bending Coupling Matrix, (B_{ij}), and the Bending Stiffness Matrix, (D_{ij}), bring together the stiffness effects from each lamina, and consequently fiber orientation, accounting for the position (z) of each lamina in the laminate (Eqs. (13)–(15)). These matrices account for the lamina thickness (t) and calculate the stress distribution based on the different loading conditions applied. A_{ij} considers the tension-compression effects of longitudinal and transverse loading, matrix D_{ij} considers the effects of bending moments, while matrix B_{ij} couples the effects of both types of loading. A relationship calculating normal forces and moments includes the above matrices as well as mid-surface strains and curvatures (Eq. (16)[1]) [10].

$$\left[A_{ij}\right] = \sum_{k=1}^{n} \left[\overline{Q}_{ij}\right]_k t_k \tag{13}$$

$$\left[B_{ij}\right] = \sum_{k=1}^{n} \left[\overline{Q}_{ij}\right]_k t_k \overline{z}_k \tag{14}$$

$$\left[D_{ij}\right] = \sum_{k=1}^{n} \left[\overline{Q}_{ij}\right]_k \left(t_k \overline{z}_k{}^2 + \frac{t_k{}^3}{12} \right) \tag{15}$$

$$\begin{Bmatrix} \hat{N} \\ \cdots \\ \hat{M} \end{Bmatrix} = \begin{pmatrix} A & \vdots & B \\ \cdots & \cdots & \cdots \\ B & \vdots & D \end{pmatrix} \begin{Bmatrix} \varepsilon^o \\ \cdots \\ \kappa \end{Bmatrix} \tag{16}$$

2.2 Failure criteria

There are three major failure modes in the microscopic level of the FRP, i.e. the constituent materials and their interface:

- Failure of the matrix phase through crack initiation and propagation.
- Failure at the reinforcing phase, which is the fracture of one or more fibers.
- Failure at the interface of the two constituents, referred to as debonding, where the fibers detach form the matrix phase.

[1] All loading conditions, including thermal and hygral effects, are accounted for in $\hat{N}$ and $\hat{M}$.

In each of the above degrade, the mechanical properties of the composite and affect the strength and the performance of the material in a different way. The fibers in FRPs, holding the load carrying capacity of the composite, constitute the phase that determines to a large degree the strength and stiffness of the material. The orientation of the fibers is crucial in stress and strain calculations, as has been previously shown through the discussion on CLT. As a result, the second failure mode, which concerns failure at the reinforcing phase, becomes of special interest, as is the one that interrupts the load carrying capacity of the fibers. It is also among the main concepts of this chapter, and will be given further attention in Section 3. Fractured fibers cannot be replaced or repaired and therefore, this failure mode permanently degrades the strength of the material.

The fracture of fibers becomes especially important when the load carrying capacity of the FRP structure is expected to be along one of the axis of the structure. Typically, fibers are chosen along or off this axis. Take for example a plate under bending. Such a plate may represent a flat beam spring (e.g. leaf springs in suspension systems) which is loaded and deflected under a bending moment. As this bending moment creates a stress distribution along the longitudinal axis of the beam, if the choice of material is FRP, the fibers are chosen parallel or at an angle to this longitudinal direction. This way the fibers hold the load carrying capacity of the beam, and the stacking sequence choice regulates the stress distribution and desired stiffness of the structure, as followed by the CLT equations. If due to failure, one or more of these fibers fracture, a discontinuity along the load carrying capacity in this longitudinal direction is generated. The specific lamina(e) with the fractured fibers become(s) responsible for the degradation of the mechanical properties of the composites, as it can no longer participate in the aforementioned CLT equations, which are exclusive to longitudinal continuous fibers.

There also exist other failure modes, such as delamination (i.e. the deboning at the lamina interface), or failure due to environmental factors (such as high moisture absorption [11, 12] or UV degradation of the matrix). In such cases, examining the failure mechanism to determine the extent to which the strength of the composite has been affected, requires investigation at the material level and its chemical composition.

Failure criteria, on the other hand, allow us to determine the effect of loading to the strength of the material. Such criteria may be used in conjunction with CLT to determine optimum stacking sequences that can guarantee a long life performance of the FRP structure at specific loading conditions before the occurrence of first ply failure.

The anisotropic nature of FRPs requires failure criteria that account for the interaction of stresses, and consequently material properties, in different directions. Such criteria are referred to as interactive failure criteria, as opposed to non-interactive ones, which focus on parameters in each direction separately (e.g. Tresca and von Mises) [6, 10, 13]. The interactive failure criteria may give a prediction of the onset of failure irrespective of the failure mode or any other conditions responsible for it (environment, thermal, etc.).

The two criteria discussed in this chapter are the Tsai-Hill and Tsai Wu. They both operate on a comparison of the stress state in each lamina to the failure stress under stress plane conditions in order to determine the failure or not of a lamina. They concern therefore, similar to CLT, the lamina level. As the majority of failure criteria, they are polynomial expansions treating the stress tensor (σ_{ij}) as the sole parameter to characterize the onset of failure. As polynomial expansions, they may be tailored to the case of transversely isotropic materials, thus reducing significantly the number of required material parameters [6]. However, because they are mere criteria, they should always be verified by experimental data, as they can only give a prediction for the onset of failure.

The stress tensor (σ_{ij}) in these criteria is calculated using CLT. As a result, it refers to the stress distribution of the FRP structure along one of the directions of the lamina. Such a lamina is considered to contain as its reinforcing phase continuous longitudinal fibers. If this is not the case, and the fibers are either discontinued or fractured, the lamina is degraded and not included in the CLT calculations, which results, in its exclusion from the following criteria (Eqs. (17) and (18)). As a result, similar to CLT the criteria presented below may provide a prediction for the onset of failure in a lamina, provided that the lamina maintains its continuous unidirectional fibers. Therefore, they would not be appropriate for failure predictions in laminae with discontinuities due to which fibers are interrupted.

Tsai-Hill Failure Criterion:

$$\frac{{\sigma_{11}}^2}{X^2} - \frac{\sigma_{11}\sigma_{22}}{X^2} + \frac{{\sigma_{22}}^2}{Y^2} + \frac{{\sigma_{12}}^2}{S^2} < 1 \tag{17}$$

In the Tsai-Hill criterion (Eq. (17)) the longitudinal (σ_{11}), transverse (σ_{22}), and shear stresses (σ_{12}) in each lamina are compared to the longitudinal tensile and compressive (X and X′), transverse tensile and compressive (Y and Y′), and shear (S) ultimate strengths. These latter strength parameters are all material parameters that may be obtained from experimental results or material databases. From the total of 5 parameters required, only 3 are involved in the equation, i.e. the above criterion becomes specific to the type of loading. If the loading results in compressive stresses, Eq. (17) will be rewritten to include the longitudinal and transverse compressive ultimate strength (X′ and Y′), as well as the shear ultimate strength. In the format presented above, it addresses failure due to tensile stresses. In either case failure has occurred when the equation on the left hand side of the criterion equals to or is greater than 1.

Tsai-Wu Failure Criterion:

$$\left(\frac{1}{X} - \frac{1}{X'}\right)\sigma_{11} + \left(\frac{1}{Y} - \frac{1}{Y'}\right)(\sigma_{22} + \sigma_{33}) + \frac{{\sigma_{11}}^2}{XX'} + \frac{1}{YY'}(\sigma_{22} + \sigma_{33})^2 + 2F_{12}\sigma_{11}(\sigma_{22} + \sigma_{33}) + \frac{1}{S'^2}\left({\sigma_{23}}^2 - \sigma_{22}\sigma_{33}\right) + \frac{1}{S^2}\left({\sigma_{12}}^2 + {\sigma_{31}}^2\right) < 1 \tag{18}$$

The Tsai-Wu criterion also investigates failure at the lamina level and states that failure occurs when Eq. (18) is equal to 1. The equation contains 6 constants involving the material parameters of tensile and compressive ultimate strengths in the longitudinal and transverse directions, as well as shear ultimate strengths. The Tsai-Wu criterion does not address failure separately due to either tensile or compressive stresses, as it includes all ultimate strengths of the material irrespective of their directionality. Additionally, it addresses stresses in direction 3, as well as shear stresses in planes including direction 3. As a result, the Tsai-Wu criterion is not exclusive to the transversely isotropic materials examined using CLT. Therefore, this criterion requires a total of 7 material parameters. The Tsai-Wu criterion terms can be evaluated by the assumption of uniaxial tension and compression results, which is based on experimental data [6, 10]. The interaction parameter (F_{12}) due to its interactive nature is an approximation that depends on the product of the products of tensile and compressive longitudinal ultimate strengths and tensile and compressive transverse ultimate strengths (Eq. (19)). It is often estimated from multiaxial stress data [6, 10].

$$F_{12} \leq \sqrt{\frac{1}{XX'} \mathrm{x} \frac{1}{YY'}} \tag{19}$$

The above strength data is obtained from experimental results on unidirectional FRPs with continuous fibers. This is one more reason, why the above criteria would fall short in accurately predicting failure in laminae with discontinuous fibers.

As mentioned above, failure criteria should be used in conjunction with experimental data for better understanding the onset of failure. Research has shown that the Tsai-Hill criterion tends to overestimate failure, while Tsai-Wu tends to underestimate failure [3–5, 14].

3. Accounting for geometric discontinuities in FRP laminates

The above discussion shows the importance of laminae fiber orientation and therefore, the stacking sequence of laminates. However, the tools discussed in Section 2, CLT and the interactive failure theories, take into account unidirectional uninterrupted fibers in the laminae. Fiber fracture is considered as one of the failure modes in FRP composites, and is one of the most detrimental ones to the material. To approach therefore, a similar analysis on structures with geometric discontinuities the above methods should be combined with further analysis tools to address the high stresses in the area of the discontinuity and avoid working with interrupted fibers.

3.1 Orienting fibers around a circular hole

To predict failure and evaluate critical stresses around geometric discontinuities in FRP laminates, different approaches and models have been developed and presented in literature. Some of these models and theories focus on fiber failure, as is for example Hashin's theory [15], while other newer approaches look into the prediction of fiber and interfiber failure [16]. In the case of geometric discontinuities, such as notches, there exist the Waddoups-Eisenmann-Kamiski (WEK) model [17, 18] that evaluates the strength of notched composite specimens using the stress intensity factor. However, the above models only evaluate failure and do not address any predictions of its onset, which is important when designing against failure.

As previously mentioned an optimum stacking sequence can improve the onset of first-ply failure in FRP laminates. As a result, the importance of an appropriate stacking sequence around a geometric discontinuity becomes even more significant. CLT has been used by Goteti and Reddy in conjunction with the stress intensity around a circular hole to examine the effect of fiber orientation, hole size, and fiber volume fraction on the stress concentration around the hole [19]. A different approach using Muskhelishvili's complex variable method and fiber orientation as input was attempted by Sharma in determining the stress concentration around circular/elliptical/triangular cutouts [20]. On the other hand, other researchers, such as Huang and Haftka, instead of focusing on the stress intensity and concentration in the discontinuity region, attempted to determine the fiber orientation around it, while keeping the fiber orientation in the remaining lamina unidirectional [21].

The aforementioned work agrees that fiber orientation around a discontinuity is affected by the following parameters:

- Size of the discontinuity (eg. diameter of a hole).
- Load type and direction

- Volume fraction of the fibers, which has already been shown to affect the mechanical properties of the FRP material as determined by Rules of Mixtures.

This chapter discusses a case study of a slightly different approach to determine fiber orientation around circular discontinuities [22]. The approach focuses on the immediate vicinity of the discontinuity, where it attempts to determine an optimum fiber orientation. It will be shown that the approach is concerned only with the plastic region around the discontinuity, where the fibers will not be interrupted, and as a result will maintain their load carrying capacity from one end of the lamina to the other. Additionally, the fiber orientation outside the plastic region of the discontinuity will remain unidirectional, based on the orientation of the lamina in the stacking sequence.

The majority of the works in literature discussing approaches to optimum fiber orientation around discontinuities or the evaluation of stress intensity in such regions use axial loading conditions. The case study presented below will assume a three point bending loading condition on the FRP laminate. In such loading cases, the majority of the aforementioned work becomes non-applicable, as the fibers in order to maintain their longitudinal load carrying capacity in the structure should remain continuous and uninterrupted. The meaning of continuous fibers, disregards the concept of fibers starting at the rim of a central to the structure discontinuity, as this considers a lamina of two sets of continuous fibers, one on each side of the discontinuity but interrupted by it.

This case study examines first CLT and the Tsai-Hill failure criterion to determine a minimum moment required to cause first ply failure in a given FRP laminate in the absence of a discontinuity, and second, the geometric stress concentration factor under bending, to determine the moment to cause failure in the presence of a circular hole. The optimum fiber orientation in the area of the hole will be determined when this minimum moment is applied. The above approach therefore, uses the aforementioned theories and criteria solely for a unidirectional lamina, and introduces linear fracture mechanics to account for the discontinuity effect.

3.1.1 Laminate beam model and discontinuity region

A six layer GFRP (Glass Fiber Reinforce Plastic) laminate with no discontinuities is being considered, at first. The laminate has a symmetric, general stacking sequence ($[0/45/0]_s$), where laminae 1 and 6 both have fibers parallel to the global horizontal dimension of the laminate (i.e. at 0°). For simplicity each layer is designed to have a thickness of 1 mm.[2]

A medium to high stiffness GFRP laminate material, S2 glass/fiber epoxy (**Table 1**), is selected with the fibers occupying 55% of the composite material volume. Each lamina is transversely isotropic, with direction 1 along the fibers and plane 23 as the isotropy plane. Therefore, information only in the 1 and 2 directions is required for CLT and Tsai-Hill calculations.

The above laminate will be compared to an identical laminate of the exact same dimensions (6 layers each at 1 mm thickness), same stacking sequence, and a central circular hole of 1 cm diameter. This second laminate will constitute the structure with the discontinuity (**Figure 2**). The loading moment M, will be determined using the Tsai-Hill failure criterion.

---

[2] GFRP laminae tend to be thinner than 1 mm. However, to simplify calculations this exaggerated thickness is chosen here.

E_1	34 GPa
E_2	8.9 GPa
G_{12}	4.5 GPa
v_{12}	0.27
X (Longitudinal tensile strength)	2000 MPa
X′ (Longitudinal compressive strength)	1240 MPa
Y (Transverse tensile strength)	49 MPa
Y′ (Transverse compressive strength)	158 MPa
S (Shear strength)	63 MPa

Table 1.
Mechanical properties of S2 glass fiber/epoxy.

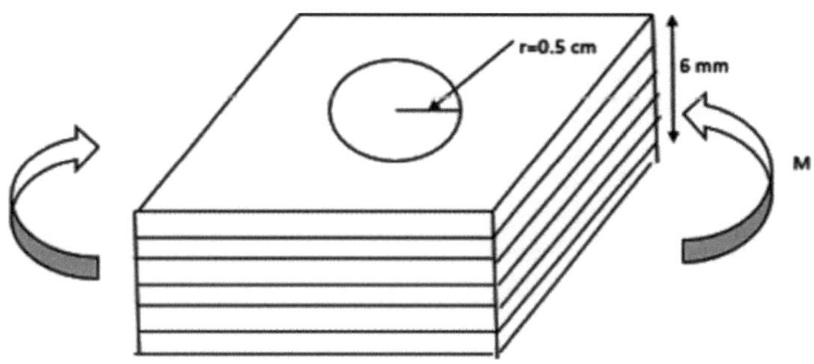

Figure 2.
Laminate with circular discontinuity of 1 cm diameter loaded by at the three point bending [22].

A plastic zone approach is used to approximate the region surrounding the geometric discontinuity that is affected by a maximum stress concentration. To determine the radius of the plastic zone (r_y) a very small crack is assumed to exist on the verge of the hole. This crack could potentially propagate in the matrix of the GFRP and lead to catastrophic failure of the matrix (one of the FRP failure modes mentioned in Section 2.2). The assumption of a very small crack allows the crack length (α) to the hole radius (r) ratio approach zero. As a result, Eq. (20) is used to estimate the Mode I (opening mode) stress intensity factor (K_I). If the stress intensity factor becomes equal to the critical stress intensity factor (K_C), the crack will begin to propagate with a radius of the plastic zone given by Eq. (21).

$$K_I = \sigma\sqrt{\pi\alpha} f(\alpha/r) \tag{20}$$

$$r_y = \frac{1}{2\pi}\left(\frac{K_c}{\sigma_y}\right)^2 \tag{21}$$

The radius of the plastic zone, as calculated in Eq. (21), begins at the crack tip. Based on the previous assumption of a very small crack length, this radius will begin on the rim of the hole, and therefore, the distance of the critical region around the discontinuity may be determined. This is the region of high stress concentration, where the reinforcement of the stacking sequence should be modified in order to strengthen the laminate. Equations (20) and (21) clearly show the effect that the size of the discontinuity has on the selection of this region.

Equation (12) in Section 2.1 calculates the stress distribution in the kth lamina of a laminate. This stress refers to a lamina of unidirectional fibers and no geometric discontinuity. Therefore, the stress concentration factor (K_t) is used to multiply this stress in order to maximize it and account for the presence and effect of a discontinuity (Eq. (22)).

$$\sigma_{max} = K_t \sigma_k \tag{22}$$

The magnitude of the stress concentration factor depends on the dimensions of the laminate, discontinuity, and loading condition. In this case study K_t = 2.7. It should be noted that the size of the hole and laminate dimensions directly affect this value.

3.1.2 Optimizing the stacking sequence around the discontinuity

Using CLT on a semi-infinite beam with no discontinuity, as the one described above, and following the analysis with the Tsai-Hill failure criterion, the minimum moment to cause first ply failure is determined. **Table 2** shows the minimum moment to cause failure in each lamina of the laminate. It can be observed from the values shown that the symmetry of the laminate and the three point bending loading conditions give a symmetry of the absolute minimum moments to cause failure in each lamina. The laminae at 0° fiber orientation are the strongest layers of the structure.

As shown in **Figure 2**, the discontinuity is at the center of the laminate and consequently each lamina, which for CLT purposes is modeled as a semi-infinite plate. Therefore, each lamina may be further modeled as symmetric in both the x and y directions of its plane, and a single quadrant of the hole may be considered for analysis. Around the quadrant of the hole a number of points are selected about which the optimum fiber orientations will be determined. The selection of these points is made based on the finite element concept of seeds around geometric discontinuities. A minimum of 16 seeds around a circular hole is recommended, and as a result, a total of 4 seeds is selected around the quadrant considered here. The analysis follows Huang and Haftka's [21] model of fiber orientation prediction, where the orientation of the fibers outside the plastic zone remains the same as the one originally prescribed to the lamina. In this study, the orientation of the fibers outside the plastic zone remains the same as that originally prescribed for the lamina in question (i.e. 0 or 45°). Depending on the accuracy required, the number of seeds around the hole may be increased. Additionally, the optimization process followed should be repeated for all laminae in the laminate, in order to provide a stacking sequence around the hole. In this case study analysis is performed for Lamina 3 at 0° fiber orientation.

Lamina/fiber orientation	Moment (Nm/m) Tsai-Hill = 1
1/0°	269
2/45°	72
3/0°	808
4/0°	808
5/45°	72
6/0°	269

Table 2.
Absolute values of minimum moment to cause failure in individual lamina.

Possible angle values in the area of the discontinuity	8°
	14°
	40°
	71°
	72°
	81°

Table 3.
Possible fiber orientations near the hole for lamina 3 at 0° fiber orientation.

Away from the plastic zone the load is carried along the unidirectional orientation of the fibers, which remains unchanged based on the given stacking sequence. However, within the plastic zone around the hole, the fiber `orientation will be constantly changing and will be following the four new orientations prescribed by the number of seeds selected. It was previously mentioned that based on symmetry, one quadrant of the plate will be considered. Therefore, the possible orientations will vary between 0 and 90°. The results obtained from the analysis may then be mirrored to the remaining quadrants in order to obtain a complete image of the fiber orientations around the hole.

$$\left(\frac{\sigma_1{}^2}{X^2} - \frac{\sigma_1\sigma_2}{X^2} + \frac{\sigma_2{}^2}{Y^2} + \frac{\tau_{12}{}^2}{S^2}\right)\rho^2 - 1 = 0 \tag{23}$$

The Tsai-Hill failure criterion is used in its polynomial form with a positive load factor (ρ) calculated as the root of the polynomial at the onset of failure, i.e. when Tsai-Hill is equal to 1 (Eq. (23)). The load factor accounts for the effect of the constantly changing fiber orientations.

A range of possible orientations near the hole for Lamina 3 are given in **Table 3**. All values are in the range from 0 to 90°. To narrow the selection of possible orientations a genetic algorithm may be applied to determine the appropriate orientations based on more specific information of the lamina and its loading. Repeating the analysis for the remaining laminae at 0° fiber orientation (laminae 1,4, and 6), is observed that that similar results are obtained, which are explained by the symmetric nature of the laminate.

4. Conclusion

FRP laminates have entered the industry world as strong and lighter material alternatives to metals, while they offer the option of an excellent material solution to many emerging technologies.

FRP laminates fail due to degradation of their mechanical properties through a range of failure modes. When designing FRPs against failure care should be taken which of the many traditional and newer approaches in predicting first-ply failure is chosen.

The stacking sequence of an FRP laminate is of great significance in determining the stress distribution in the laminate as well as predicting first ply failure. Using CLT and interactive failure criteria an optimum stacking sequence may be determined for specified loading conditions, or the loads to cause first ply failure can be calculated when the stacking sequence of the laminate is known. However, the above theories and criteria are limited to addressing unidirectional and continuous laminates with no geometric discontinuities.

To perform a similar analysis on FRP laminate structures with holes or other geometric discontinuities, the above methods should be combined with other techniques or models that account for the presence of a discontinuity. In this chapter a case study is used to show such an approach in an attempt to determine the stacking sequence around a circular hole. The limitations of CLT and the interactive failure criteria are overcome with the use of fracture mechanics and more specifically the concepts of stress intensity and stress concentration factors. The approach uses CLT and the Tsai-Hill criterion to predict the loads and lamina of first ply failure, and then fracture mechanics to determine a plastic zone around the discontinuity and maximize the stresses in this region. As a result, a multitude of new possible fiber orientations are calculated, which can be used as the extension of the lamina fiber orientation around the hole to strengthen the lamina in that region and prevent or delay failure, without interrupting the fibers and consequently the load carrying capacity of the FRP.

Appendices and nomenclature

E_{ijkl}	Young's Modulus
G_{ij}	shear modulus
K	bulk modulus
f	volume fraction
v_{ij}	Poisson's ratio
σ_{ij}	stress tensor
ε_{kl}	strain tensor
$\boldsymbol{\kappa_{ij}}$	curvature
$\boldsymbol{\varepsilon_{ij}^{o}}$	mid-surface strains
$\boldsymbol{\alpha_{ij}}$	coefficient of thermal expansion
$\boldsymbol{\beta_{ij}}$	hygroscopic coefficient
$\overline{Q}$	stiffness matrix.
A_{ij}	extension stiffness matrix
B_{ij}	extension-bending coupling matrix
D_{ij}	bending stiffness matrix
z	position of layer in laminate
X and X′	longitudinal tensile and compressive strength
Y and Y′	transverse tensile and compressive strength
S	shear strength
K_I	mode I stress intensity factor
K_C	critical stress intensity factor
K_t	stress concentration factor
α	crack length
r	hole radius
r_y	plastic zone radius
σ_y	applied yield stress
ρ	function of the orientations around the discontinuity

Author details

Roselita Fragoudakis
Merrimack College, North Andover, USA

*Address all correspondence to: fragoudakisr@merrimack.edu

References

[1] Gürdal Z, Haftka RT, Hajela P. Design and Optimization of Laminated Composite Materials. 1st ed. New York, NY: Wiley-Interscience; 1999

[2] Suresh S. Fatigue of Materials. Cambridge, UK: Cambridge University Press; 1998

[3] Fragoudakis R, Saigal A. Predicting the fatigue life in steel and glass fibre reinforced plastics using damage models. Journal of Materials Science and Applications. 2011;**2**:596-604

[4] Fragoudakis R, Saigal A. Using damage models to predict fatigue in steel and glass fibre reinforced plastics. Journal of Materials Science and Engineering With Advanced Technologies. 2011;**3**:53-65

[5] Fragoudakis R. In: Aly A, editor. Failure Concepts in Fiber Reinforced Plastics, Failure Analysis and Prevention. Croatia: Intech; 2018. ISBN: 978-953-51-5230-9

[6] Christensen RM. Mechanics of Composite Materials. Mineola, NY: Dover; 2005

[7] Barbero EJ. Introduction to Composite Materials Design. 2nd ed. Boca Raton, FL: CRC Press; 2010

[8] Dvorak G. Micromechanics of Composite Materials. New York: Springer; 2013

[9] Vasiliev V, Morozov EV. Advanced Mechanics of Composite Materials and Structural Elements. 3rd ed. UK: Elsevier; 2013

[10] Staab GH. Laminar Composites. Woburn, MA: Butterworth-Heinemann; 1999

[11] Dhakal HN et al. Effect of water absorption on the mechanical properties of hemp fibre reinforced unsaturated polyester composites. Composites Science and Technology. 2007;**67**: 1674-1683

[12] Wang W. Study of moisture absorption in natural fiber plastic composites. Composites Science and Technology. 2006;**66**:379-386

[13] Jones RM. Mechanics of Composite Materials. 2nd ed. New York, NY: Taylor & Francis, Inc.; 1999

[14] Fragoudakis R. Predicting the optimum stacking sequence of fiber reinforced plastic laminated beams under bending. In: SAMPE Seattle 2017; 22–24 May 2017; Seattle, WA; 2017

[15] Hashin Z. Failure for unidirectional fiber composites. Journal of Applied Mechanics. 1980;**47**:329-334

[16] Ribeiro MC et al. Finite element analysis of low velocity impact on thin composite disks. International Journal of Composite Materials. 2013;**3**:59-70

[17] Waddoups ME, Eisenmann JR, Kaminski BE. Macroscopic fracture mechanics of advanced composite materials. Journal of Composite Materials. 1971;**5**:446-451

[18] Kannan VK, Rajadurai A, Nageswara Rao BN. Residual strength of laminated composite after impact. Journal of Composite Materials. 2010; **45**:1031-1043

[19] Goteti C, Reddy S. Influence of fiber volume fraction, fiber angle and hole size in the stress concentration around the circular hole of an orthotropic lamina under unidirectional in plane loading. International Journal of Applied Science & Engineering. 2014;**2**:1-12

[20] Sharma D. Stress concentration around circular/elliptical/triangular

cutouts in infinite composite plate. In: Proceedings of the World Congress on Engineering, III; 2011

[21] Huang J, Haftka RT. Optimization of fiber orientations near a hole for increased load- carrying capacity of composite laminates. Structural and Multidisciplinary Optimization. 2005; **30**:335-341

[22] Fragoudakis R. A numerical approach to determine fiber orientations around geometric discontinuities in designing against failure of GFRP laminates. International Journal of Structural Integrity. 2019;**10**: 371-379. DOI: 10.1108/IJSI-10-2018-0064

Chapter 2

Propagating Stress-Strain Curve Variability in Multi-Material Problems: Temperature-Dependent Material Tests to Plasticity Models to Structural Failure Predictions

Vicente Romero, Amalia Black, George Orient and Bonnie Antoun

Abstract

This chapter presents a practical methodology for characterizing and propagating the effects of temperature-dependent material strength and failure-criteria variability to structural model predictions. The application involves a cylindrical canister ("can") heated and pressurized to failure. Temperature dependence and material sample-to-sample stochastic variability are inferred from very limited experimental data of a few replicate uniaxial tension tests at each of seven temperatures spanning the 800°C temperature excursion experienced by the can, for each of several stainless steel alloys that make up the can. The load-displacement curves from the material tests are used to determine effective temperature-dependent stress-strain relationships in ductile-metal plasticity models used in can-level model predictions. Particularly challenging aspects of the problem are the appropriate inference, representation, and propagation of temperature dependence and material stochastic variability from just a few experimental data curves at a few temperatures (as sparse discrete realizations or samples from a random field of temperature-dependent stress-strain behavior), for multiple such materials involved in the problem. Currently unique methods are demonstrated that are relatively simple and effective.

Keywords: materials, modeling, calibration, uncertainty, thermal-structural failure

1. Introduction

Sandia National Laboratories is developing the capability to adequately model the complex multiphysics leading to pressurization and breach of sealed compartments that contain organic materials such as foams, which volatilize when the compartments are heated in fire accidents. The present chapter along with references [1–7] describes aspects of the associated activities, including experiments, modeling and simulation, code and calculation verification, and advanced model validation and uncertainty quantification (UQ) methods.

The modeling and verification, validation, and uncertainty quantification (VVUQ) activities were performed under a multiyear "abnormal thermal-mechanical breach" (T-M breach) task [1] of a Predictive Capability Assessment Project (PCAP) in the Verification & Validation (V&V) subelement of the U.S. Dept. of Energy Advanced Simulation and Computing (ASC) program. The goal of the PCAP T-M breach task was to assess the error and quantify the uncertainty in modeling the thermal-chemical-mechanical response and weld-related breach failure of sealed canisters ("cans") weakened by high temperatures and pressurized by heat-induced pyrolysis of foam. The planned outcome of the PCAP T-M breach task was to measure improvements in prediction accuracy over time as the models and computer platforms became more capable.

The Sandia Weapon System Engineering and Assessment Technology Campaign (WSEAT) program supported the project by conducting material characterization tests and validation experiments [2] (see **Figure 1**). This partnership provided an opportunity to develop a fully integrated process from design of experiments through model validation assessment, with uncertainty reduced as much as possible and propagated through the process.

Breach failures were expected to occur, and in the tests, they did occur, at the circumferential perimeter (laser) weld that joins the top lid to the can sidewalls. This is because the weld thickness is significantly less than the can lid and sidewalls (see **Figure 2**), and the tests/cans of interest in this chapter were heated at the lid top surface, so the top weld material was much hotter/weaker than the perimeter weld material at the bottom of the can. While prediction of canister internal temperatures, time to breach, and breach pressure are sought in the T-M breach task, breach pressure is the quantity of interest (QOI) in this chapter.

This chapter describes a practical methodology for characterizing and propagating the effects of variability of material strength and failure criteria to structural response and failure predictions involving multiple temperature-dependent materials. Relatively simple and effective UQ techniques are used to model and propagate temperature dependence and material sample-to-sample variability effects inferred from very limited material characterization tests.

Figure 1.
Thermal-chemical-mechanical validation experiments [2], including internal pressure response. The 'can' includes the cylindrical 'sidewalls' or 'walls,' as well as the top 'lid' and bottom 'base.'

Figure 2.
Close-up of modeled geometry where can top lid, sidewall, and internal foam meet. (Nominal geometry values are 0.03 in. weld depth, 0.0645 in. wall thickness, and 0.007 in. clearance between the lid and sidewall in the weld region.)

Section 2 summarizes the material characterization tests and results. These involve uniaxial tension tests on several cylinder specimens at each of seven temperatures spanning the 800°C temperature excursion experienced by the can, for two stainless steel alloys that make up the can and weld materials. Section 3 summarizes the ductile-metal material constitutive models used for elastic-to-plastic stress-strain response at a given temperature. The procedure to parameterize the constitutive model's stress-strain relationships through inverse analysis to best match measured load-deflection data curves from the tension tests is also explained and demonstrated. Section 4 describes the material damage models and failure criteria calibrated to the experimental stress-strain data. The thermal-chemical-mechanical models for predicting can thermal, pressurization, and structural response (and failure) are also briefly summarized. Section 5 describes the use of the models and associated simulations to propagate effects of material strength and failure variability to estimate breach failure pressure variability. Sensitivity analysis is also performed to assess the relative contributions of the various materials' strength and failure criteria variability on the total variability of predicted failure pressure. Section 6 provides some summary observations and conclusions.

2. Temperature-dependent material strength characterization tests and results

Round-bar tensile tests were conducted at seven temperatures: 20, 100, 200, 400, 600, 700, and 800°C for both the can lid and base (bar) material and the sleeve/wall (tubular) material. Most specimens were in the axial orientation (see **Figures 3** and **5**), but some tests for the lid material were conducted in the radial orientation at 20, 600, and 800°C to provide an indication of orientation dependence.

2.1 Tensile characterization for PCAP 304L stainless steel lid and base material

Round-bar tensile-test specimens of 0.3 in. (inch) diameter and extensometer gage length 0.80 in. were used in the tension tests described here. Specimens in the axial direction were extracted from 3.5 in. diameter bar stock as shown in **Figure 3**. The can top lids and bottom bases were machined from this lot of bar stock.

Figure 3.
Axial tensile specimen extraction from bar stock.

Material strength stress-strain characterization tests (uniaxial tensile tests) were conducted [8] on an MTS 880 20 Kip axial test frame with displacement (stroke) control to produce a nominal strain rate of 0.001/s. This strain rate was based on the model-predicted conditions in the PCAP thermal-mechanical breach experiments [2]. A strain rate of 0.0001/s, based on computed local strain rates in the weld region, was also tested to explore the sensitivity of the material to strain rate. However, strain-rate effects were not included in the PCAP material strength model because it was figured that in the accident scenarios being assessed here, strain-rate effects were of secondary importance prior to reaching a stress-strain maximum load condition where our failure criteria would be activated (see Section 4). For conditions past maximum load, it is well known that 304L stainless steel (ss) has nonnegligible strain-rate dependence at all temperatures.

The test results for the PCAP lid material in the axial direction are shown in **Figure 4** in terms of engineering stress versus engineering strain. As expected, the strength of the material decreases as the temperature increases. However, around

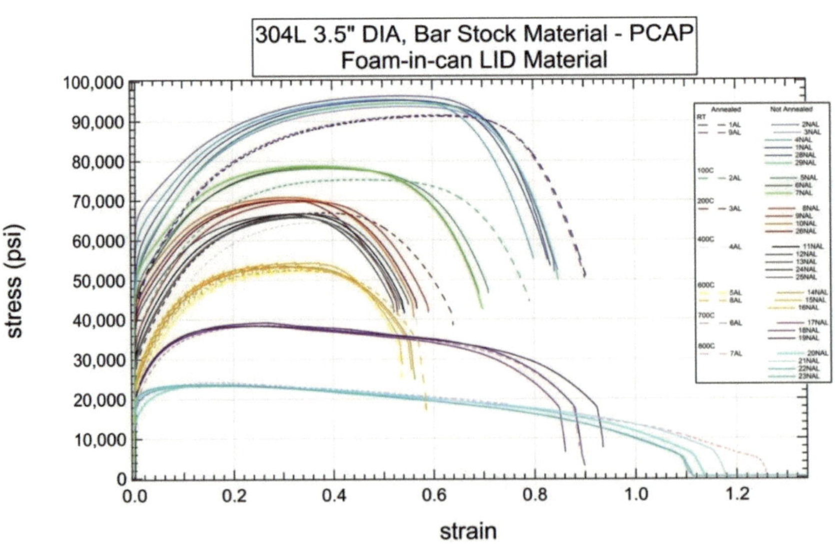

Figure 4.
Engineering stress vs. engineering strain curves for PCAP lid and base material (ss 304L, axial specimens from bar stock).

600°C, there is a noticeable inflection point in the temperature related shape trend of the stress-strain curves. It is believed that this inflection occurs because the deformation mechanisms change from void growth and deformation to grain slippage at about half of the material melt temperature for 304L stainless steel (see last subsection of Section 2). Half the melt temperature of 304L stainless steel is roughly 700°C.

Reannealed and not re-annealed sets of specimens were tested to better quantify the effect of the material starting condition on the tensile properties. It was presumed that annealing the tensile specimens would not have a large effect since the raw materials were reported to be in an annealed condition, but the test results do show a noticeable difference [1], which seems to indicate that the original material was not in a fully annealed state. The specimens that were reannealed in this test were placed in a vacuum at 1000°C for 30 min. The effect of reannealing the specimens was larger at the lower temperatures. As the test temperature increased, this effect became less noticeable, and by 700°C, it is fairly indistinguishable.

Specimens cut from the cylindrical bar stock in a direction normal to its axis were tested at 20, 600, and 800°C to provide an indication of orientation dependence. These results are not shown, but typically exhibited ultimate stress values at lower test displacements/strains and with sharper subsequent weakening than the axial samples did show. Nonetheless, an isotropic constitutive model was used (see Section 3.1). It was calibrated with stress-strain curves from tension tests with the axial-cut specimens only.

2.2 Tensile characterization for PCAP 304L stainless steel wall material

Specimens in the axial direction were extracted from the tube-stock material as shown in **Figure 5**. The can sleeve/sidewalls were machined from this stock. The

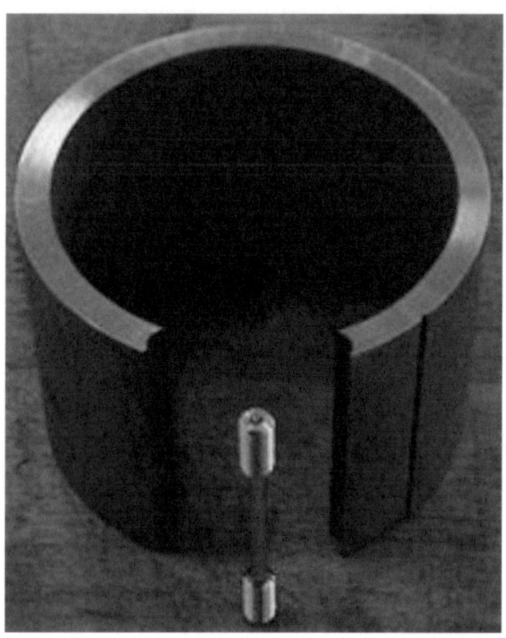

Figure 5.
Axial tensile specimen extraction from wall material.

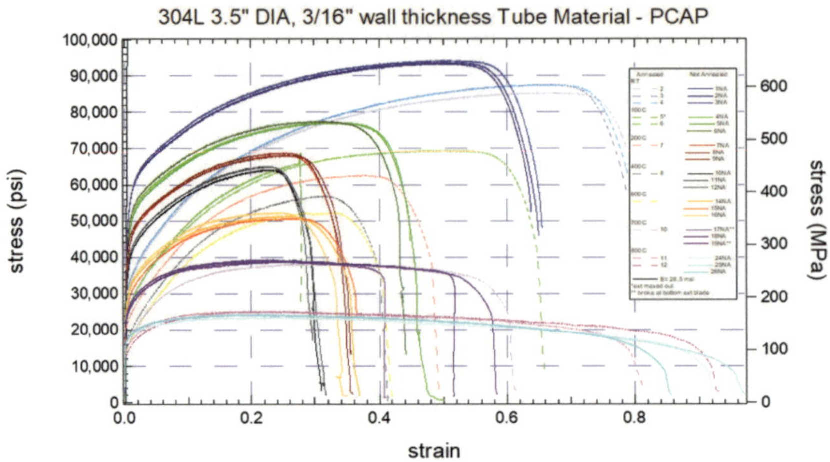

Figure 6.
Engineering stress vs. engineering strain curves for PCAP tube (wall) material (ss 304L, axial specimens from tube stock).

nominal specimen diameter was 0.1 in. with an extensometer gage length of 0.62 in. The test results for the PCAP wall material in the axial direction are shown in **Figure 6**. Like the lid material, the strength of the wall material decreases as the temperature increases and a noticeable change in the temperature related shape trend of the stress-strain curves occurs at about 600°C.

2.3 Ignored but possible creep and strain-rate effects

In general, the engineering stress versus engineering strain curves for both the wall and lid materials exhibit a markedly different character above 600°C. Below this temperature, the ultimate strain decreases as the temperature increases, but above 600°C, features of the stress-strain curve change and the ultimate strain becomes larger. Considering that test data are only available at one strain rate, a plausible explanation includes the hypothesis that at about half of the melt temperature, creep deformation is observed manifesting in creep relaxation in a displacement-controlled tensile test (see [9]).

For the PCAP test conditions, a temperature of 600°C translates to a homologous temperature of 0.48. The yield stress at 600°C is about 25 ksi. Converting yield stress to shear stress and then normalizing it by the shear modulus (76.3 GPa) gives a value of 1.1e−3. These conditions are right at the transition to power-law creep [9], which is where dislocations are able to climb (through thermal fluctuations) over precipitates and other barriers. Thus, a creep-dependent model may be necessary for temperature conditions above 600°C.

Additional factors may include temperature and strain-rate–dependent phase transformation mechanisms as reported in [10]. At room temperature, a martensitic phase appears when strain is loaded. However, this does not appear to happen at elevated temperatures leaving the material in an austenitic phase, which is weaker than martensitic phase, allowing more necking at elevated temperatures.

These potential mechanisms still need to be further investigated. A creep and strain-rate–dependent phenomenological constitutive model is recommended in future studies with elevated temperatures, especially if the material deformation will be simulated past maximum load conditions.

3. Temperature-dependent material strength characterization tests and results

Mechanical constitutive behavior was modeled using a strain-rate-independent isotropic ductile-metal multilinear elastic plastic (MLEP) plasticity model (e.g., [11]). The parameterized form of the "true" stress-strain curve (Cauchy stress-plastic logarithmic strain) consistent with the constitutive model's formulation is represented in piecewise linear fashion by multiple linear segments as described in Section 3.2. Fundamental assumptions of the constitutive model follow.

3.1 Constitutive model description

Strain-rate independence: Load-displacement response of a specimen subjected to constant strain rate is independent of the strain-rate magnitude. For ductile metals and over the small strain rates and small range of rates encountered in the PCAP application, this is considered an acceptable assumption below about half of the melting temperature. However, it is known that 304L stainless steel exhibits some strain-rate dependence at lower temperatures as well, especially past a maximum load condition. In the PCAP project, this issue was examined by testing specimens at several strain rates expected to represent the bulk strain rates of the cans in the tests. This was used to assess model-form uncertainty since local strain rates are expected to spatially vary over the regions surrounding stress concentrations, like at the welds. Ultimately, strain-rate effects were judged to be small relative to other modeling errors and uncertainties in the PCAP T-M breach problem.

Independence on hydrostatic stress: Independence of yield behavior on the hydrostatic stress state in metals is well understood. Plastic deformation is attributed to shear states of stress and strain characterized by the second invariant of the deviatoric stress tensor (J2). Definitions of the von Mises effective stress σ_eff and the equivalent plastic strain (EQPS) are derived from this statement [13]. The effective stress and equivalent strain are uniaxial measures that allow collapsing triaxial stress and strain states in structural applications into uniaxial measures that map into experimental stress-strain curves derived from uniaxial load-displacement tests.

Isotropy: The inelastic constitutive response is independent of the load orientation. This is a reasonable assumption for metal components unless they have been fabricated with processes that introduce directional character of the grain structure such as rolling or forging without annealing.

The MLEP model is a standard metal plasticity constitutive representation for industry practice. MLEP treatment helps FE models to be affordable with reasonable computational resources and is suitable as long as the limitations are understood and not violated significantly. The MLEP model only relates stress and strain; no intrinsic statement about material strength-related failure is made. Material failure modeling is discussed in Section 4.1.

3.2 Constitutive model parameterization procedure

Material characterization involves solving an inverse problem to determine the MLEP constitutive model's "true" Cauchy stress-plastic logarithmic strain relationship that recovers the load-displacement or engineering stress-strain data (e.g., **Figure 4**) from tensile tests. As such, a fitting procedure was used to enable the inverse calculation.

Before the onset of necking, the true stress and true strain in a tensile specimen can be calculated from the load-displacement recorded from the load cell of the testing frame and an extensometer mounted on the specimen. Once necking occurs, the true strain in the middle of the necked region must be calculated from a finite element (FE) model of the gage section of the specimen. The ASC massively parallel solid-mechanics code Adagio [12] was used for the simulations. To ensure that necking initiates between the ends of the gage section, a small imperfection is introduced in the mesh. Section 3.3 investigates hex-mesh density sufficiency.

The implicit relationship between the load-displacement response of the FE model and the MLEP constitutive relationship necessitates implementing an iterative procedure to fit an MLEP model to the load-displacement record obtained from testing. Since the test data contain a large number of potentially noisy data points, some data conditioning through down-sampling and/or smoothing is necessary, resulting in order 20 data points. This is based on engineering judgment and is ultimately confirmed by comparing the load-displacement curve with the entire test record. Point selection could be done on the experimental data record or by conditioning the experimental data and selecting points from the smooth conditioned data.

Assuming that the multilinear true stress-true strain is fitted at the ith point, the next linear section is obtained by a two-step process illustrated in **Figure 7** and summarized next.

1. Bracket the slope of the next segment.

    a. Extend the current MLEP curve by the current slope candidate to the next strain point in the conditioned dataset.

    b. Solve the FE model with the current candidate MLEP model loaded with the strain point.

    c. Evaluate the reaction force in the gage section. If the force is less in the conditioned dataset, decrease the slope candidate and repeat 1.a; otherwise, slope candidates bracketing the actual slope are found.

2. Solve for the slope resulting in a reaction force that matches the force on the experimental dataset. In the current implementation of the MLEP fitting method, the bisection algorithm is used.

This process requires careful management of analysis restarts to efficiently iterate on the MLEP line segment slopes. Once all the line segments are determined, an analysis is run with the complete MLEP line segment set to characterize necking through the entire strain history.

Noise in the test data was sometimes a factor in the MLEP calibrations for this project, especially at high temperatures. Cubic spline smoothing [14] was used to smooth test data when needed. The raw and smoothed data are illustrated in **Figure 8**. The view is closely zoomed to a section of the curve.

It was visually observed that the differences between MLEP curves calibrated to different specimens (specimen-to-specimen differences) were much larger than the fitting errors from the calibration process itself.

Several factors influence the success of the iterative MLEP procedure:

Model gage-length and cross-section mismatch to test specimen geometry: This may make the model too stiff to follow necking behavior. Iteration usually fails near the ultimate load, resulting in a true stress-true strain curve with constant slope.

Element refinement: When the mesh is coarse, the model is too stiff to produce the target necking behavior. For strains beyond ultimate load, the iteration fails, resulting in a true stress-true strain curve with constant slope.

Load step: It may be necessary to reduce the load step in the solid-mechanics simulation so that the calibration procedure does not miss the necking behavior, which would drop the final engineering stress-strain curve below the test data.

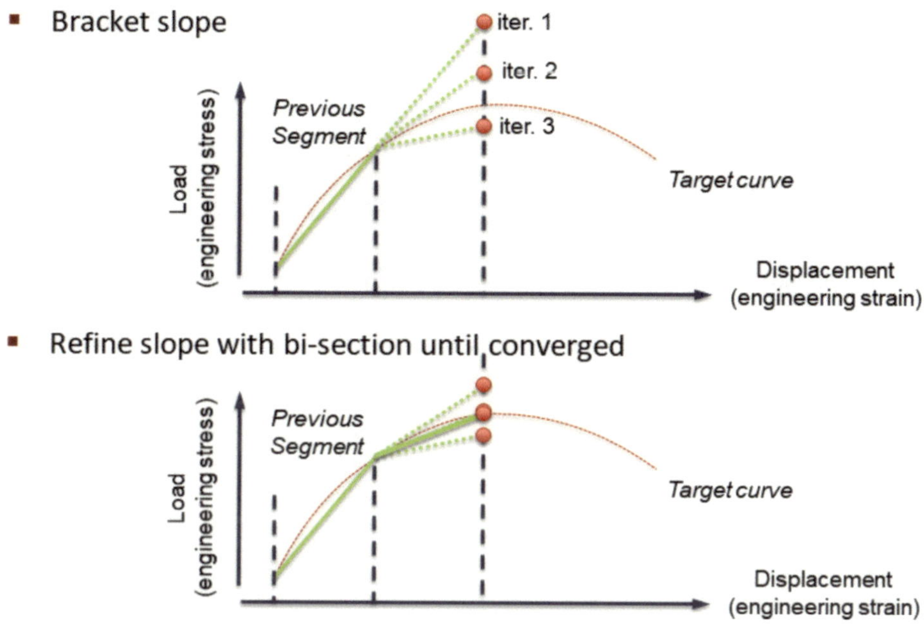

Figure 7.
Iteration process to arrive at next point in piecewise linear parameterization of MLEP stress-strain curve.

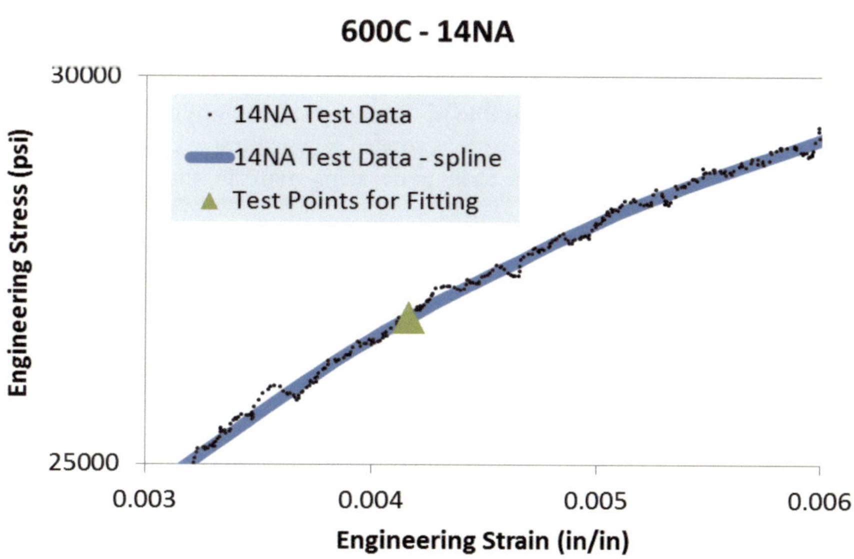

Figure 8.
Cubic-spline smoothing of tension test data

Number of points before and after ultimate load: Similar to the load step effect, too few points result in the analysis taking the wrong path.

Noisy test data: An increased level of noise was observed in the data at elevated temperatures. This may actually be a result of a physical phenomenon but the modeling approach in this study assumes a smooth stress-strain behavior, and constitutive curves with abrupt changes in slope may throw the iteration off course. Data conditioning with piecewise smooth regression fit usually addresses this adequately.

Stress and slope tolerance: Insufficient tolerance in the bisection method may result in failure to converge (and, therefore, constant slope true stress-true strain curve) caused by unstable undershooting/overshooting.

Confinement of necking within the extensometer gage length: It is possible that numerically the model necks at the ends of the gage section instead of the middle. Always check the final necking pattern before accepting the fitted model.

Amount of mesh imperfection to induce necking in the middle of the gage section: This is an artificial imperfection, and care must be taken not to reduce the cross-section area significantly. Typically, 0.1% or less artificial reduction of area is desired although this guideline may not be sufficient for "flat" stress-strain curves. Insensitivity with respect to this numerical uncertainty needs to be demonstrated.

Element type selection: Necking tends to excite the hourglass modes, if present, and the hardening curve might be altered by low resistance to shear deformation in the necked section.

The calibration process requires the elastic modulus, yield stress, and Poisson's ratio as input. While the experimental load-displacement curves have data at small strains (<0.2%), they have been measured with extensometers optimized for large strains (>50%). No specifications regarding their accuracy at small strains were received, and it was observed that while the lid data showed an initial linear section consistent with literature data for elastic modulus and yield stress, the wall data exhibited no significant linear section. The decision was made to use the modulus/yield data obtained from the test record for the lid and literature data for wall.

The MLEP model is not parametric in the sense of a power law or a Johnson-Cook constitutive model, where a handful of parameters describe the shape of the true stress-true strain curves. A cubic spline elastic-plastic (CSEP [15]) version of MLEP now exists where the true stress-strain curve is parameterized by order 7 stress-strain points or knots whose stress and strain values are simultaneously optimized such that the experimental load-displacement or engineering stress-strain curve is best matched. This appears to also generally work well but often requires more model runs. In either case, generating material-curve data fits is not easy.

3.3 Solution verification

The MLEP fit process has been used by analysts at Sandia National Laboratories to fit room-temperature curves for several years and experience has shown that about 16 elements across the radius of a cylindrical gage section have been sufficient. High-temperature 304L curves, however, exhibit rather "flat" characteristics and low yield points, so the concern of using mesh-converged models was revisited here. The results in **Figure 9** show that, for the final number of elements (32) used in the FE simulations in the MLEP calibrations, the calibration results are well converged up to and beyond the maximum load point where the stress-strain curves will be evaluated in the can-level simulations (because of the material failure criteria being tied to the maximum load point; see Section 4.1).

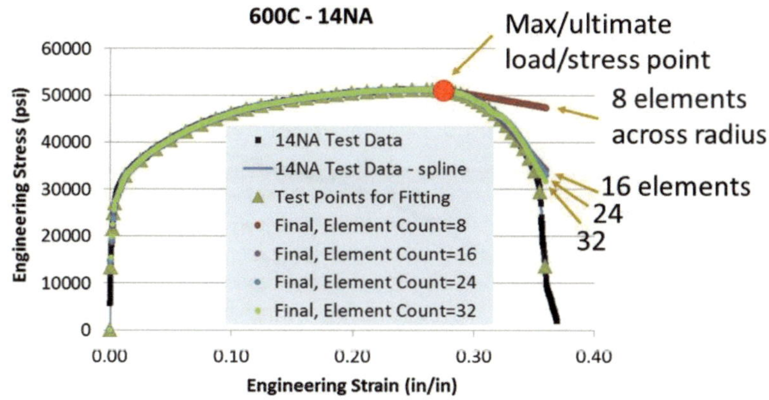

Figure 9.
Solution verification for gage-section FE model and solves used for MLEP model calibrations.

3.4 Selected calibration results

Calibration to a set of wall specimen test data at room temperature is shown in **Figures 10** and **11**. The yellow symbols indicate the calibrated curve, the test data are shown with black symbols, and the blue symbols identify the points selected from the spline smoothed data where the MLEP fit was performed. **Figure 11** illustrates the following: (a) the fact that the load-displacement record is not linear at small strains; (b) so literature data were used for modulus and yield stress (second blue marker); and (c) the MLEP calibration iteration is stopped when convergence criteria are satisfied; the knee of the yellow curve and the second blue marker (the target) do not coincide exactly. Convergence criteria were decided based on considerations of the expected impact on QOIs and practical computational throughput limits.

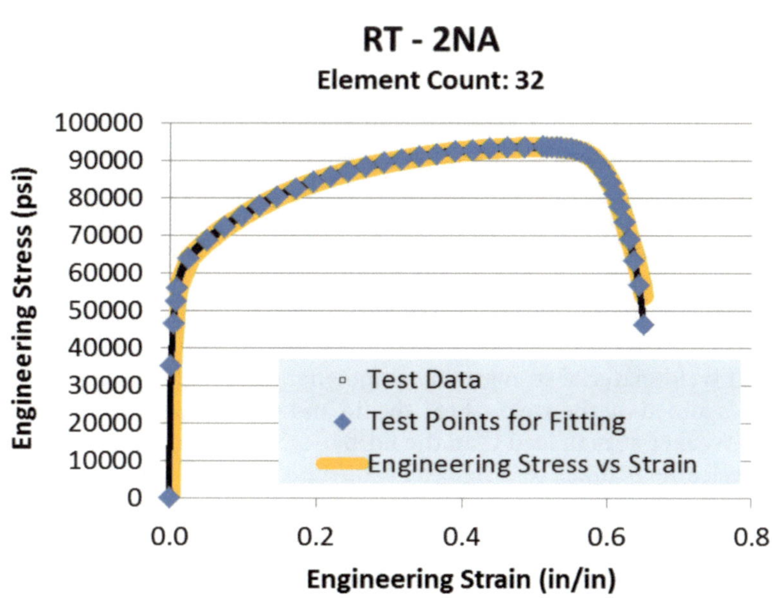

Figure 10.
Room temperature MLEP calibration.

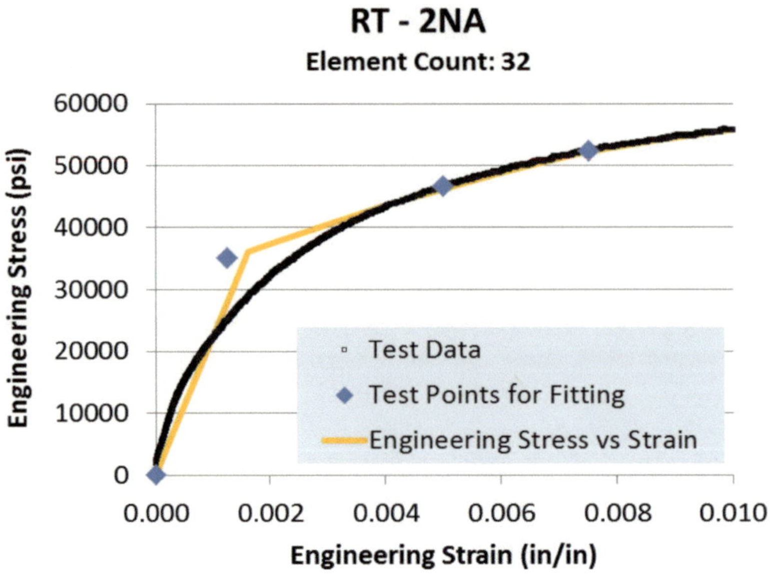

Figure 11.
Room temperature calibration, small strains.

3.5 Process automation and archiving

Considering the number of calibration instances (49 sets of test data comprised of several replicate tension tests at seven temperatures for two materials), the calibration process was automated in a script, and the different instances of calibration were executed under the control of a DAKOTA [16] parametric study.

4. Material failure criteria and can pressurization and response/failure modeling

4.1 Weld material modeling and failure criteria

It was originally planned to obtain weld material stress-strain curves and failure criteria by calibrating to tension tests of butt-weld square bar specimens and then validating to can pie-section weld flexure tests to failure. However, both endeavors proved to be problematic experimentally and computationally [1] such that adequate model accuracy could not be established.

As a reasonable alternative, the following approach was taken. For welds of normal quality that do not have anomalies like voids, empirical evidence strongly suggests that weld material strength lies somewhere between the strengths of the two materials joined by the weld—here the lid and the can wall. Wall/tube material was slightly weaker at max load than the lid/bar-stock material, so we made a conservative-leaning choice to assign the wall/tube material curves and failure criteria to the weld.

Microstructural examination of the PCAP cans pressurized to failure indicated ductile overload failure of the laser welds of the heated lids. Equivalent plastic strain (EQPS) and tearing parameter (TP) are candidate models for accumulated material damage, as explained next. These models' computed damage values at the point of maximum load and engineering stress in the uniaxial tension tests are taken to be critical material failure criteria for these two models. This is consistent with current failure modeling practice at Sandia National Laboratories in conjunction with MLEP models in overload failure modes. Because of the notorious difficulty of predicting structural failure from material damage modeling, the two models and their failure criteria were used and assessed as candidate indicators of *onset* of failure in the PCAP application.

As paraphrased from [11], the TP failure indicator within Sierra uses an approach based on the work of Brozzo et al. [17]. This parameter takes the form of an evolution integral of the stress state integrated over the plastic strain. Two modifications were needed beyond Brozzo's original formulation. The first modification was the inclusion of a Heaviside bracket on the maximum principal stress. That is, if the maximum principal stress is compressive (negative), the increment to the tearing parameter is zero. Thus, there is no increase in material "damage" for compressive stress states, nor is there "damage healing" for compressive states.

The second modification to TP calculation resulted from the investigation of notched tensile test results. Two sets of notched round bar tensile tests referenced and summarized in [11] were performed on different heating treatments of 6061-T6. Comparison between the simulations and the experimental results showed excessive ductility for the simulations using the original formula. By raising the stress-state portion of the integral to the fourth power, a match between experimental data and simulation results was achieved. The final form follows. It is used as well for the ductile stainless steels in the present work.

$$TP = \int \left\langle \frac{2\sigma_T}{3(\sigma_T - \sigma_m)} \right\rangle^4 d\varepsilon_p \tag{1}$$

In this equation, σ_T is the maximum principal stress; σ_m is the mean stress (average of principal stresses); and ε_p is the equivalent plastic strain, EQPS.

All input parameters, including the critical value of TP that coincides with material "failure" as interpreted below can be obtained from a model calibrated to a standard tensile test as explained below. The mechanical response code with MLEP constitutive model requires a Cauchy stress and plastic logarithmic strain to define strain-hardening behavior. To determine this from a standard tensile test, it is necessary to solve the inverse problem described in Section 3.2. More details of the tearing parameter model for ductile failure can be found in [11].

In addition to the critical TP value, the equivalent plastic strain critical value is also used to predict failure in order to assess the uncertainty due to failure-model form. EQPS is derived directly from the strain (e.g., [13]).

Although TP and EQPS critical values are often defined based on tension test material separation failure, the critical values for this project were defined at the maximum load in the tension tests. This decision was made for two reasons. First, the global loading of the can structure is due to pressurization, and the pressure is always increasing and will cause incipient failure when a maximum load condition is reached. Second, the weld failures observed in the can tests showed little evidence of necking. Up to maximum load, there is little necking. It was reasoned that defining critical failure values based on any finite element in the model reaching the hardening curve maximum load point identified from the tension tests would result in conservative failure predictions for the can.

Therefore, the failure criteria defined at the maximum load in the tube/wall round bar tension tests were used to signify weld material failure in the can breach predictions. **Figure 12** shows the location (i.e., circle) of the max load condition used to obtain the critical values for EQPS and for TP at each temperature from the stress-strain curves. The critical values were obtained by calibrating the MLEP

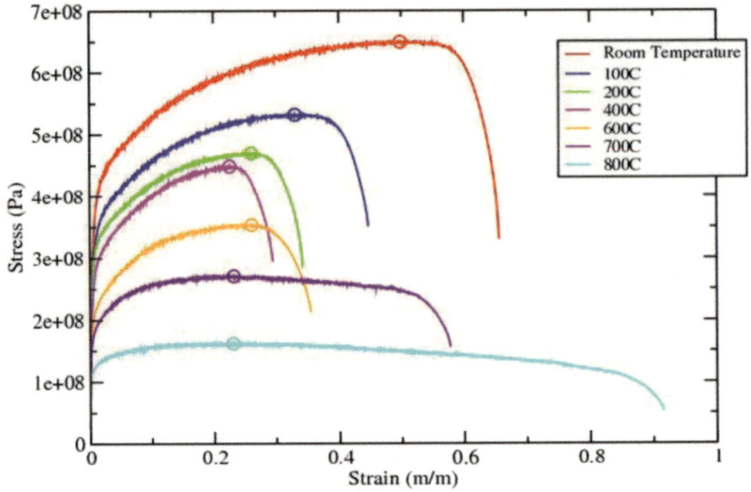

Figure 12.
Maximum load locations (circled) where critical failure values of TP and EQPS are obtained (from ss 304L tube/wall material tension tests, only one curve is shown at each temperature for illustration).

TEST #	TEMP (°C)	Critical TP	Critical EQPS
1NA	20	0.40411	0.400321
2NA	20	0.419455	0.415633
3NA	20	0.403365	0.399622
4NA	100	0.303277	0.300511
5NA	100	0.293452	0.290511
6NA	100	0.276675	0.273797
7NA	200	0.224306	0.221862
8NA	200	0.236491	0.233974
9NA	200	0.225990	0.223562
10NA	400	0.203274	0.201021
11NA	400	0.205928	0.203682
12NA	400	0.206650	0.204383
14NA	600	0.237258	0.234747
15NA	600	0.272458	0.269665
16NA	600	0.217022	0.215917
17NA	700	0.184131	0.182011
18NA	700	0.208394	0.206040
19NA	700	0.234210	0.231570
24NA	800	0.233182	0.230782
25NA	800	0.193157	0.192174
26NA	800	0.164383	0.162645

Table 1.
Max load-related failure criteria values for TP and EQPS determined from tension-test specimen model calibrated to each ss 304L tube/wall experimental load-displacement curve.

model to match the load-displacement curves (engineering stress-strain curves) from a mesh-converged model of the specimen gage section and then by searching for the maximum TP and EQPS values on the specimen midsection. There were replicate tension tests at each temperature and the corresponding critical values were determined for each data curve as listed in **Table 1**.

4.2 Models for can thermal-chemical-structural response and failure

The thermal-chemical-mechanical models used are briefly summarized here from [1]. The Sandia SIERRA module [18] for massively parallel thermal-fluid computations was used to model the heating of the can, its thermal response, and thermally-induced chemical-kinetic decomposition of the foam [19] and resulting gas species generation that causes pressurization. The solid mechanics and structural modeling module [12] were used to model the mechanical response of the can and failure at the weld under pressurization and high temperatures and large temperature variations in time and space. The module uses a nonlinear quasi-statics finite element approach based on a Lagrangian, three-dimensional, implicit scheme. A multilevel iterative solver enables solution of problems with large deformations, nonlinear material behavior, and contact. Temperature-dependent elasto-plastic constitutive models are accommodated, where the elastic parameters (Young's modulus, Poisson's ratio, and yield stress) and the stress-strain plasticity curves are temperature dependent.

The thermal-chemical simulation provides the temperature and pressure boundary conditions for the mechanical model. The only feedback from the mechanical model to the thermal-chemical model is the can's internal volume change due to deformation. The volume change affects the pressure level in the can through the Ideal Gas Law, which is evaluated within the thermal module and then communicated to the mechanical module. The can geometry is not changed/updated in the computational heat-transfer model because the can deformation is fairly slight (lateral bulging equivalent to a few can-wall widths) so it is thought to negligibly affect the heat transfer (or at least not affect the heat transfer in the model, given the way it was modeled). The heat transfer and foam decomposition submodels and parameters are also not affected by pressure in the current treatment. (The uncertainties associated with including pressure effects on these phenomena were judged larger than the error involved by not including pressure effects, and any modeling error effects would be quantified through the validation comparisons [1, 4] that were the culmination of the PCAP assessments).

The thermal-chemical and mechanical models were run in a "concurrent but segregated" manner in which Sandia's SIERRA [20] software framework for massively parallel multiphysics computations passed temperature, pressure, and volume information between the thermal-chemical simulation and the mechanical simulation. SIERRA coordinates and manages the different time-stepping of the thermal-chemical and mechanical codes and the transfer of spatial temperature fields solved on the tetrahedral thermal mesh to nodal temperature assignments to the nodes of the mechanical hex mesh.

The full 360-degree can geometry with internal foam was used for the thermal-chemical simulations, and the 90-degree pie-slice geometry without foam was used for the mechanical simulations. The full 360-degree geometry was used in the thermal-chemical simulations because at the time, the foam and enclosure radiation models did not accommodate any kind of symmetry boundary conditions.
The mechanical simulations were much more computationally expensive, so a

quarter-can partial geometry without foam was used to reduce cost. Leaving foam out of the mechanical model tremendously reduces the number of finite elements and thus computational cost, and is thought to have negligible impact on structural behavior and pressure-breach failure in the PCAP problem.

In the thermal model, a uniform heat flux boundary condition was applied on the lid surface. The flux level was calculated as follows to be consistent with the temperature data from the experiment control TCs. The four control TCs were fully inserted into radially drilled holes at midplane on the lids at 0, 90, 180, and 270 degrees around the lids [2]. A proportional-integral-derivative (PID) routine [21] was used to determine the heat flux magnitude needed to match the control thermocouple temperature responses. This approach results in a more realistic temperature distribution versus using a TC-guided uniform temperature condition over the entire lid surface.

On the side walls and base of the can, convection and radiation boundary conditions were specified (as described in [5]) to represent the heat transfer between the can exterior and the surrounding environment.

Different element types and mesh densities are used as appropriate in the thermal and mechanical models. Code verification activities were performed for the thermal and solid/structural mechanics codes and models [1]. For the order-200 thermal-mechanical and mechanical-only simulations run for VVUQ and sensitivity analysis in the PCAP project, an affordable mesh size of 1.85 million hex elements for the structural model (12 elements through the thickness of the weld) and 14.3 million tet elements for the more affordable thermal model were used. This affordable 'Level 4' mesh was one in a succession that went up to Level 6 with approximately double the number of elements in the structural and thermal models (see [3]). The succession of meshes was used for a solution verification assessment in [1] to estimate and account for Mesh 4–related error/uncertainty in the VVUQ analysis and results in [4]. Solver tolerances were experimented with and set to contribute small error/uncertainty relative to mesh effects.

Figure 13 shows the Level 4 mesh at a critical portion of the structural model where weld failure is determined in the thermal-structural simulations. Stress

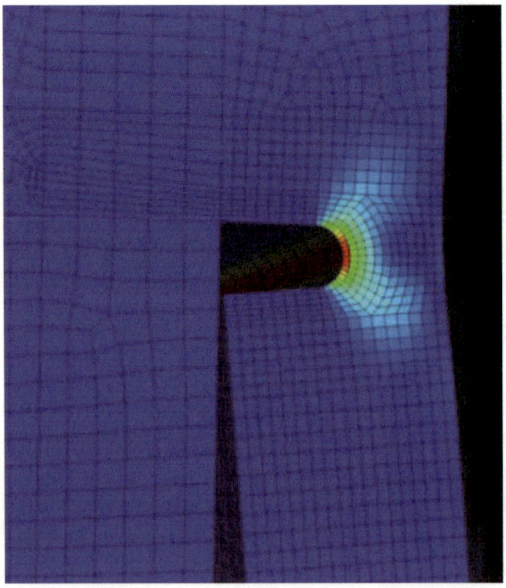

Figure 13.
Weld-section close-up of structural-model Level 4 hex mesh used in model validation, UQ, and sensitivity analysis simulations in the PCAP project. Stress concentration is evident at the crown of the weld notch. (Figure from [3].)

concentration is evident at the crown of the weld notch. This type of weld-geometry representation was found to best support weld failure predictions analyzing many different geometry representation schemes [1, 22].

5. Discrete propagation of material strength and failure variability to can breach-pressure variability predictions

The material strength uncertainty sources treated here come in *discrete* (not parametric) form of multiple slightly varying stress-strain curves and failure criteria representing stochastic material strength variations in the can lid, weld, and wall materials. These curve-to-curve variations and failure criteria when propagated cause predicted variability (and uncertainty thereof) in can response and failure pressure level. The 16 uncertainties not related to material strength and failure variability are all parametric in nature and are held at nominal values listed in [4] for the purposes of the following material-curve propagations and analysis of results.

Sensitivity studies in [1, 4] of the effects of the more prominent modeling uncertainties regarding thermal, pressurization, and structural phenomena in the PCAP T-M breach problem reveal that material curve strength variations are among the most significant causes of failure-pressure predicted variability and uncertainty thereof.

5.1 Dealing with temperature dependence of the material stress-strain curves

Dealing with temperature dependence of the material curves adds a significant difficulty to the discrete propagation problem. This is addressed in the following two data processing steps before propagation can be performed in Section 5.2.

Step I: material stress-strain curves strength-to-failure ranking and down-selection

In this step, the effective strength of the repeat material curves at each temperature was ranked and then down-selected to three representative curves (high, medium, and low strength) according to predicted failure pressure predictions from the PCAP can simulations. The curve-strength ranking process at a given temperature is much more involved when multiple materials exist than when only one material exists (which allows a simple straightforward process, [23]). This is because the strength ranking of a given set of material curves can depend on the particular combination of material curves used for the other two materials (e.g., wall strength and flexure can affect stress-strain phenomena at the weld notch). There are many such combinations because each of the two other materials has multiple material curves, so the ranking investigation should involve confirming curve ranking is robust over all or at least a few different test combinations of the other materials' curves. This is addressed in the rather involved and computationally expensive ranking process, summarized in the Appendix.

Step II: correlation and Interpolation of stress-strain curves across temperatures

This step proceeds from a precedent in [23] and Step I's determinations, portrayed conceptually in **Figure 14**. Three material curves of low, medium, and high effective strength exist per characterization temperature. When several material curves exist at each temperature, for UQ purposes, strength is assumed to be highly correlated across temperatures such that a curve with higher relative

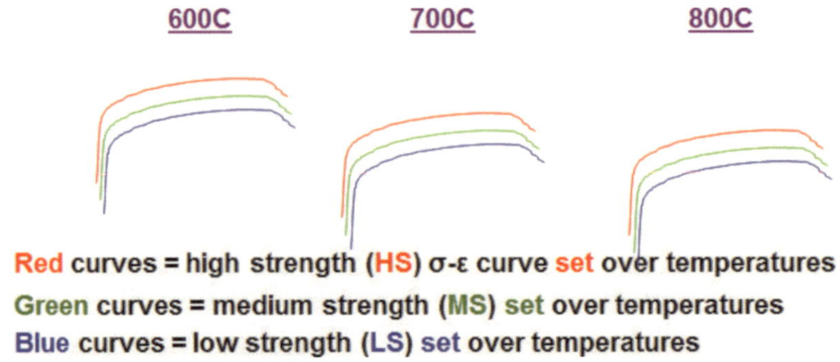

Figure 14.
Notional portrayal of high, medium, and low effective strength stress-strain curves at adjacent characterization temperatures.

strength at lower temperatures is assumed to retain higher relative strength at higher temperatures. This assumes that material weakening mechanisms and % weakening are roughly similar with increasing temperature whether the material is initially of higher, medium, or lower relative strength.

The correlation assumption appears physically reasonable and tremendously reduces the number of potential combinations of material curves to be sampled when a material transitions temperatures. For example, there are 3 × 3 × 3 = 27 potential combinations of material curve combinations in the figure that could be used in a simulation that transitions temperatures from 600°C to 700°C to 800°C. So to investigate all these potential combinations would take 27 simulations with the expensive PCAP can model. To transition all seven temperatures would present 37 = 2187 possible combinations. This is just for one material. For the three materials in this problem, each with three material curve options, this would present 21,873 ≈ 1010 possible combinations. Clearly, this is unaffordable and seems wholly unnecessary given the reasonableness of the temperature-strength correlation assumption.

Hence, for each material, we link, for example, its high-strength curves across the seven characterization temperatures. We interpolate across the characterization temperatures as follows: at a temperature in-between two adjacent characterization temperatures, the stress is linearly interpolated from the stress values (at the applicable input stain level) from the two stress-strain curves at the upper and lower enveloping temperatures.

For each material, this effectively gives one constructed high-strength, temperature-varying, stress-strain function. Temperature-dependent medium and low strength functions are likewise constructed. For this problem, we end up with

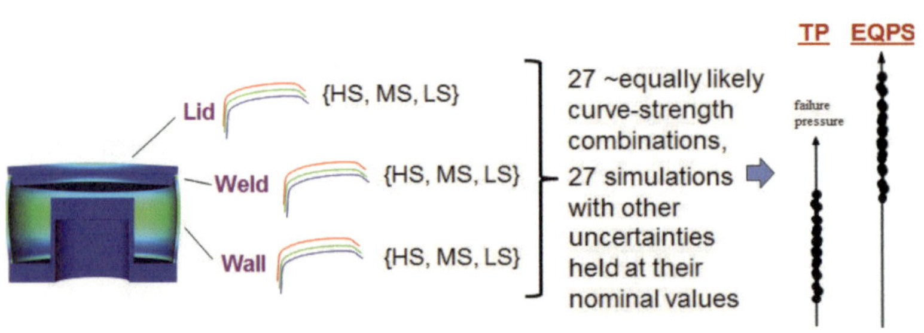

Figure 15.
Notional depiction of PCAP can materials' high strength (HS), medium strength (MS), and low strength (LS) temperature-dependent stress-strain functions, and propagation of the material strength variability via 27 assumed equally-likely combinations of material strength functions (e.g., one combination is lid MS function/ weld HS function/wall LS function).

each material having high strength (HS), medium strength (MS), and low strength (LS) temperature-dependent stress-strain functions as depicted in **Figure 15**.

5.2 Stress-strain function uncertainty propagation, results, and sensitivities

Given the constructed high, medium, and low strength stress-strain functions for the materials, a strategy in [1, 4] was taken to form and propagate all 27 possible combinations of stress-strain functions as conveyed in **Figure 15**. The model (Mesh 4) was run with experimental heating and other conditions summarized in [5] from Test 6 in [2]. This is the reference nominal test of the five replicate tests in the PCAP validation assessment (see [4]). This yields 27 failure pressures for each of the TP and the EQPS failure criteria as depicted in the figure.

5.2.1 Sensitivity analysis

The 27 failure pressures for the TP and EQPS failure criteria are plotted in **Figure 16**. The left columns of the TP and EQPS results are for the HS stress-strain function for the weld material coupled with nine different (all possible) combinations of lid and tube/wall strengths varied over their LS, MS, and HS options. Similarly, the center and right columns of results in the figure are, respectively, for

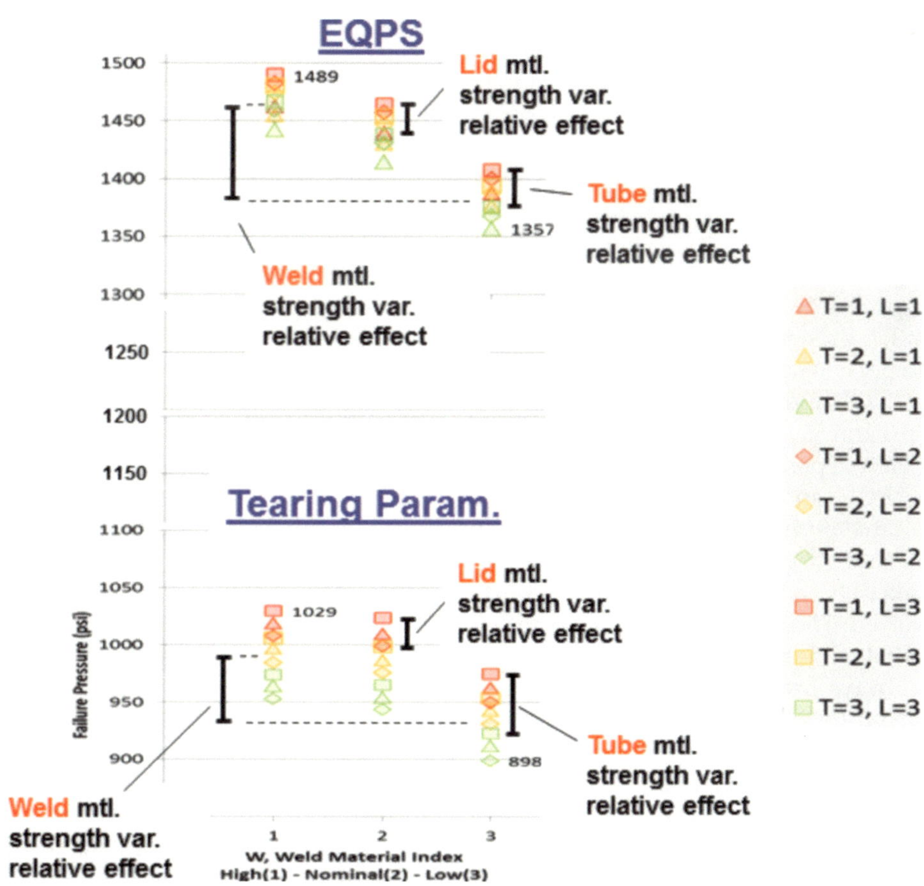

Figure 16.
Predicted failure pressures and sensitivities for different combinations of lid, weld, and tube material strength functions.

MS and LS weld strength functions coupled with the nine possible combinations of lid and tube/wall strengths.

The EQPS results are an average of about 460 psi or 50% higher than the TP failure pressures. (It was later determined that much of this difference could be explainable by very underconverged Mesh 4 results with the EQPS damage model; see [3, 4].) For both failure criteria, the individual and average failure pressures decrease as expected from column to column as the weld material strength decreases from HS to MS to LS. The decreases are somewhat greater with the EQPS failure criterion than with the TP criterion. This is reflected in the relative sizes of the interval bars labeled *weld material strength variation relative effect* in the figure. These mark the average decrease in failure pressure when weld strength goes from high to low. For EQPS, the interval bar has a length of 83 psi = average of HS weld column (1468 psi) minus average of LS weld column (1385 psi). For TP, the interval bar has a length of 55 psi = average of HS weld column (993 psi) minus average of LS weld column (938 psi).

For EQPS, the vertical ordering of results within a column does not change from column to column as weld strength decreases from HS to MS to LS. The ordering of TP results is also consistent across columns but is slightly different from the ordering of EQPS results, as discussed below. Both TP and EQPS results within a given column (where weld strength is fixed) are marked by a symbol shape that identifies the lid material strength (like the triangle that corresponds to a lid strength 1 = High per the legend at the bottom of the figure). The color of the symbols signifies the strength level for the tube/wall material: red, yellow, and green correspond to 1 (High), 2 (Medium or Nominal), and 3 (Low) strengths.

The predicted failure pressures in **Figure 16** always show that the red instance of a given symbol corresponds to a higher failure pressure than the yellow instance, which is always higher than the green instance. Thus, for any given weld and lid material strength combination, the predicted failure pressure increases with wall strength. The interval bars labeled *tube material strength variation relative effect* in the figure signify a representative magnitude of increase in failure pressure when tube/wall strength rank goes from low to high (green to red for a given symbol shape). For EQPS, the plotted interval bar has a representative magnitude of 32 psi. For TP, the corresponding interval bar has a much larger tube/wall strength effect of representative magnitude of 52 psi. As expected, these tube/wall strength effects are less than the weld strength EQPS and TP average failure pressure effects of 83 psi and 55 psi.

Failure pressure orderings relative to lid material strength rankings are less intuitive. EQPS and TP results within a given column (where weld strength is fixed) are indicated by a given symbol shape while holding the symbol color (signifying tube/wall strength) fixed. A stronger lid might make the weld fail at lower pressure than a weaker lid because there is more bending occurring at the weld if a stiffer lid is involved. This proposition is supported by the TP results ordering that the triangle symbols (lid strength 1 = High) and the diamond symbols (lid strength 2 = Medium) are always lower than the rectangle symbols (lid strength 3 = Low). However, the proposition conflicts with the TP ordering of the diamond symbols (lid strength 2 = Medium) always being lower than the triangle symbols (lid strength 1 = High). This apparent nonmonotonic behavior of predicted failure pressure with lid-strength could be due to the nonisothermal can temperatures underlying the simulation results here, whereas isothermal cans were used to rank the lid curve strengths. This could indicate that the isothermal lid curve-strength ranking process was not fully robust.

For EQPS, results within a given column and for a given color show a monotonic ordering fully consistent with the said proposition: the rectangle symbols (lid strength 3 = Low) are always highest on the plot and then the diamond symbols next (lid strength 2 = Medium), with the triangle symbol (lid strength 1 = High) always lowest.

The sizes of the interval bars labeled *lid material strength variation relative effect* in the figure are indifferent to any potential curve strength ordering errors because all 27 curve combinations are used. For both TP and EQPS, the said interval bars have representative magnitudes of 25 psi each. As expected, these lid strength variation effects are smaller in magnitude than, but not insignificant relative to, the effects of weld and tube/wall material strength variations.

Thus, for both TP and EQPS failure criteria, weld material strength variations have the largest effect, as expected, but lid and wall strength variations also have significant effects.

5.2.2 Uncertainty processing and interpretation of failure pressure results

We now consider the uncertainty processing and interpretation of the pressure failure results. If dealing with multiple but few stress-strain curves for only one material, then appropriate uncertainty treatment has been established and confirmed in the series of papers and reports [24–27]. The approach recognizes that the stress-strain curves are discrete realizations with no readily identifiable parametric relationship between them. Yet, the stress-strain curves come from and belong to a larger population that reflects the material's variability. Fortunately, a mathematical description of the generating function for the larger population of ss curves is not needed with the approach summarized next. The output scalar data (the predicted failure pressures) are worked with, rather than attempting to create a parametric or spectral generator function that is consistent with the ss curve data realizations.

In analogy with **Figure 15**, an application of the approach with, say, three stress-strain function curves for a single material would result in three predicted failure pressures with the EQPS or TP failure criteria. (We could also work with other scalar output responses of interest, like displacement, strain, or Von Mises stress at a given point on the can and at a given time, or even spatial-temporal maxima as scalar quantities that vary with the three input stress-strain function curves.) Because only three function curves and corresponding failure pressure realizations exist, small-sample related error will typically exist in any characterization of aleatory uncertainty due to the stochastic material strength variability. Thus, substantial small-sample epistemic uncertainty exists concerning the error in characterizing the aleatory variability.

A small number of realizations or samples will usually underpredict the true variance of material strength and related failure pressures or other responses. Mean or central response will also usually be significantly mispredicted. Potential significant nonconservative small-sample bias error can result, causing unsafe engineering design and risk analysis, even if the physics prediction model was perfect in every other way.

Statistical tolerance intervals (TIs) attempt to compensate for sparse sample data by appropriately biasing response estimates. For instance, the three failure pressure values would be processed into 95%coverage/90%confidence TIs (95/90 TIs). With reasonably high reliability, these estimate conservative but not overly conservative bounds on the "central" 95% of response from very sparse random samples/realizations of the input data. The central 95% of response is the range between the 2.5 and 97.5 percentiles of the true response distribution that would arise from an infinite number of samples. This central 95% range has been found to be convenient and meaningful for model validation comparisons of experimental and model-predicted aleatory response quantities (e.g., [23, 28–30]), which is also the purpose [4] of the present UQ results.

Investigations in [26, 31, 32] have concluded 95/90 TIs to be preferable to many other UQ methods tried or critically assessed for estimating, from very sparse sample data, conservative but not overly conservative bounds on the central 95% of response. The other methods tried or critically evaluated include bootstrapping [33], optimized four-parameter Johnson-family distribution fit to the response samples [34], nonparametric kernel density estimation specifically designed for sparse data [35], nonparametric cubic spline PDF fit to the data based on maximum likelihood [36], and Bayesian sparse-data approaches [37].

The TI approach is also much easier to use than the other UQ methods investigated. A 95/90 TI is constructed by simply multiplying the calculated standard deviation $\tilde{\sigma}$ of the data samples by a factor f to create an interval of total length $2f\tilde{\sigma}$. The interval is centered about the calculated mean ($\tilde{\mu}$) of the samples.

The multiplying factor f is readily available from tables in statistical texts (e.g., [38, 39]), formulas (e.g., [40]), or software (e.g., [41]) that encodes the formulas. The factor is parameterized by two user-prescribed levels: one for the desired "coverage" proportion of stochastic response, and one for the desired degree of statistical "confidence" in covering or bounding at least that proportion. For instance, a 95%coverage/90%confidence TI prescribes lower and upper values of a range that is said to have at least 90% odds that it spans at least 95% of the true probability distribution from which the random samples were drawn. The said 90% odds or confidence exist only when sampling from a Normal distribution. Reduced confidence levels for non-Normal distributions are discussed next.

Although derived for Normal populations, 95/90 TIs will span the central 95% ranges of many other sparsely sampled PDF types with reasonable/useful odds or confidence. For instance, 89% of 144 PDFs (including highly skewed and multi-modal highly non-Normal distributions) studied in [25–27] had empirical confidence levels of 75% or greater with 95/90 TIs and N = 4 random samples. From studies in [26] on several representative PDFs, it is projected that 90% of the 144 PDFs would have confidence levels > 85% with 95/95 TIs and N = 4.[1] These average or expected confidence levels decline slowly as the number of samples increases.

Although TIs often provide reliably conservative estimates, TIs can egregiously exaggerate the true variability when very few samples are involved. This is a downside that comes with high confidence levels of bounding the true central 95% of response.

Now, we consider the problem where the output response samples come from discrete stress-strain function variations of "multiple" materials as in the present problem. A naive approach would be to construct (e.g., 95/90) TIs from the 27 failure pressure values indicated in **Figure 15** for the TP and EQPS failure criteria. However, TIs pertain to random sampling of the contributing input uncertainties, where for 27 response samples, each of the contributing source uncertainties would typically be sampled at 27 different values. Repeat values would not ordinarily occur, especially with a moderately small number of samples like 27. This is not the case here; each input stress-strain function of a given material is sampled repeatedly

[1] Confidence levels of 75% or 85% are often adequate to manage risk, especially if conservatism from other sources exists in the analysis or results—such as when several sources of uncertainty are present where each involves sparse data conservatively treated with the TI method. Studies in [32] and [42] indicate that when more than one dominant or influential uncertainty sources are sparsely sampled and represented conservatively with TI confidence levels of say >70%, when the conservatively represented uncertainties are combined in linear propagation or aggregation, the individual conservative biases compound to yield substantially greater than 70% confidence of conservative bias in the combined uncertainty estimate.

(nine times) in the course of propagating all possible combinations of curves. So it was decided that constructing TIs using N = 27 would not be appropriate. (This was later confirmed by studies on a linear test problem in [42].) Instead, because only nine independent realizations of input information exist in this problem (three stress-strain functions for each of three materials), it was ventured that TIs should be constructed based on an effective number of samples N = 9. Having no more-fundamental basis to proceed on at the time, this course was taken in the PCAP VVUQ project [1, 3–7].

There is a lack of well-established sampling methods for identifying combinations of model inputs from sparse quantized sets of choices or "levels" in the various factors (i.e., the levels are not prescribable; the few available stress-strain functions are the only "levels" available), such that propagation of the relatively small number of affordable or available input combinations will yield appropriate response statistics and distribution information. Subsequent to the PCAP project, investigations in [43] provide a more fundamentally grounded approach. For the present problem, it would construct and average TIs based on failure pressure results from propagating selected sets of the 27 possible combinations of material curves according to an analogy with Latin hypercube sampling (LHS [44]) as explained next.

5.2.2.1 Latin hypercube sampling analogue for discrete material curves, and associated TIs

Latin hypercube sampling of one or more continuous input random variables is well recognized as an efficient sampling method for Monte Carlo propagation of probabilistic uncertainty through general nonlinear response functions or models (e.g., [45, 46]). With LHS and continuous random variables, M samples of a given *output* variable corresponds to M points in a D-dimensional space of D input random variables, where each input random variable is sampled at M different values or realizations of that variable. An analogue of this type of treatment exists for our discrete random function UQ problem as follows:

a. For each of the three materials, there are M = 3 different realizations of stress-strain functions.

b. For each material, choose one strength level, for example, form an input data combination {weld HS, tube/wall LS, lid LS} (refer to **Figure 15**). This is one of the possible 27 combinations of the materials' stress-strain functions discussed previously.

c. Run the model with these input curves to predict a corresponding failure pressure.

d. Do this three times; each time create a new random combination of input curves that does not use a curve that was previously selected and used. This yields three simulations, each with a single curve from each material, where each material curve is used once and only once over the prediction set of three simulations.

e. The three failure pressures predicted from the three simulations are used to construct a 95/90 TI based on n = 3 samples of response, in analogy with n = 3 TI that would be constructed for three samples of response from LHS MC with continuous random variable inputs.

5.2.2.2 Averaging equally legitimate TIs to reduce chances of extreme/ nonrepresentative TIs

This methodology yields a prediction set of M = 3 results from M = 3 LHS combinations of the 27 possible combinations. Many such equally legitimate M = 3 prediction sets can be formed. **Table 2** lists three equally legitimate sets as examples. Across any row for Set 1, 2, or 3, the high-strength, medium-strength, and low-strength function curves appear and are used once and only once for each of the materials (Weld, Tube/Wall, and Lid).

Each set in the table leads to a legitimate TI. There is no apparent reason to favor one LHS input set and TI result over another. Therefore, one could think about equally weighting the various TI results to get an "average TI" by averaging the individual TI upper ends to get an average TI upper end, and similarly to get an average TI bottom end. The average TI might be better than the individual TIs in that the average TI has a reduced chance of being an anomalous nonrepresentative result from an extreme/nonrepresentative sample set that could be obtained by random chance. A constraint on the averaging strategy is that the averaged TIs should ideally come from LHS sets that are *diverse* as a group. This means they do not have input-sample combinations (so output response samples) in common between the sets. **Table 3** shows three individually legitimate LHS sets that are nondiverse as a group because all sets have a response sample based on the same input Combination A and corresponding output response sample.

Diverse TI averaging was performed on a "Can Crush" solid-mechanics UQ test problem where reference truth results were available as part of the development of the test problem [43]. The test problem had two material variability sources each with two aleatory realizations of stress-strain curves. It was found that TI-averaging improved 95/90 TI success rates of bounding the true central 95% of response by 9 percentage points over the average success rate of individual TIs. A success rate of 94% was obtained for average TIs over a test matrix of 16 output quantities and 10s of random trials for each quantity. Individual TIs had a lesser but still reasonable success rate of 85% on average over the same 16 quantities. Similar results have been found on a second solid mechanics test problem with a different constitutive model and structural failure problem. This is now in the process of being written up.

Example 3-run LHS sets	Combination A {weld, tube/wall, lid}	Combination B {weld, tube/wall, lid}	Combination C {weld, tube/wall, lid}
Set 1	{H,H,H}	{M,L,M}	{L,M,L}
Set 2	{L,L,L}	{M,H,M}	{H,M,H}
Set 3	{L,M,H}	{M,L,L}	{H,H,M}

Table 2.
Three diverse LHS sets of material curves combinations.

Example 3-run LHS sets	Combination A {weld, tube/wall, lid}	Combination B {weld, tube/wall, lid}	Combination C {weld, tube/wall, lid}
Set 1	{H,H,H}	{M,L,M}	{L,M,L}
Set 2	{H,H,H}	{M,M,L}	{L,L,M}
Set 3	{H,H,H}	{M,L,L}	{L,M,M}

Table 3.
Three LHS sets of material curves combinations that are nondiverse *between sets.*

Note that the TI averaging method does not require more experimental data (realizations of stress-strain random functions), but does require more model runs. For the present PCAP UQ application, one legitimate TI requires three model runs. To average three equally legitimate TIs results would require nine model runs total. It appears that averaging three or four individual TIs represents the knee in the cost vs. reliability curve (the size of confidence intervals on sample means has a sharp knee at 4 samples). Therefore, TI averaging incurs $\sim 3\times$ computational cost over producing a single TI, but for the regime of solid mechanics UQ problems discussed in this chapter, the total computational cost would often be limited to ~ 10 model runs. This is a moderate computational cost to pay for the likely significant improvement from TI averaging. It is also small relative to the computational cost of the material curve ranking procedure that temperature dependence requires. The computational cost would also usually be small compared to the cost of getting the experimental data in the first place, or getting more of it.

In related methodology, reference [6] presents a method for aggregating the aleatory uncertainty of response (failure pressure) from propagated discrete aleatory realizations of functional data (per the present chapter), with aleatory uncertainty from propagated parametric random variables. Ref. [7] demonstrates how to further handle, in a practical way, any epistemic parametric uncertainty that may be involved in the UQ problem.

Finally, if the model predictions are to be used to support estimation of small "tail" probabilities of response for robust/reliable design or safety/risk analysis, the sample results from the LHS sets in 2 would be processed in a different way. This is demonstrated in recent investigations in [26, 47–49] on 16 diversely shaped distributions and tail probability magnitudes from 10^{-5} to 10^{-1}. Reliably conservative and efficient estimates of small tail probabilities are obtained. Further reliability and accuracy benefits occur from averaging multiple estimates from equally legitimate subsets of samples from the available sparse-data pool (i.e., from use of statistical jackknifing).

6. Conclusions

This chapter presented a practical and reasonable methodology for characterizing and propagating the effects of temperature dependent material strength and failure-criteria variability to structural model predictions. Particularly challenging aspects of the application problem in this chapter (and often in other real applications) are the appropriate inference, representation, and propagation of temperature dependence and material stochastic variability from just a few experimental stress-strain curves at a few temperatures (as sparse discrete realizations or samples from a random field of temperature-dependent stress-strain behavior), for multiple such materials involved in the problem. Currently unique methods are demonstrated that are relatively simple and effective. The practical methodology is versatile and flexible for application to other solid-mechanics problems involving constitutive model calibration to sparse functional temperature- and/or strain-rate-dependent data, and then propagation of the incorporated uncertainty to application models and their output quantities.

Authour note

Sandia National Laboratories is a multi-mission laboratory managed and operated by National Technology and Engineering Solutions of Sandia, LLC., a wholly owned subsidiary of Honeywell International, Inc., for the U.S. Department of Energy's National Nuclear Security Administration under contract DE-NA0003525. This manuscript describes objective technical results and analysis. Any subjective views or opinions that might be expressed in the paper do not necessarily represent the views of the U.S. Department of Energy or the United States Government.

Appendix: Process for stress-strain curves strength-to-failure ranking and down-selection

Table A1 lists the material tests for the bar stock (can lid and base) and tube stock (can sidewall and weld) characterized at seven temperatures (Section 2). The weld material is specified to be the same as the wall material because it was weaker than the lid material, so it provides a more conservative representation of the weld

	20°C	100°C	200°C	400°C	600°C	700°C	800°C
Lid—1	2NAL	5NAL	8NAL	11NAL	14NAL	17NAL	20NAL
Lid—2	3NAL	6NAL	9NAL	12NAL	15NAL	18NAL	21NAL
Lid—3	4NAL	7NAL	10NAL	13NAL	16NAL	19NAL	22NAL
Lid—4	28NAL	—	26NAL	24NAL	—	—	23NAL
Lid—5	29NAL	—	—	25NAL	—	—	—
Weld—1	1NA	4NA	7NA	10NA	14NA	17NA	24NA
Weld—2	2NA	5NA	8NA	11NA	15NA	18NA	25NA
Weld—3	3NA	6NA	9NA	12NA	16NA	19NA	26NA
Tube—1	1NA	4NA	7NA	10NA	14NA	17NA	24NA
Tube—2	2NA	5NA	8NA	11NA	15NA	18NA	25NA
Tube—3	3NA	6NA	9NA	12NA	16NA	19NA	26NA

Table A1.
List of all material curves for the lid, weld, and tube (wall) at each temperature.

	20°C	100°C	200°C	400°C	600°C	700°C	800°C
Lid—1	2NAL	5NAL	8NAL	11NAL	14NAL	17NAL	20NAL
Lid—2	3NAL	6NAL	9NAL	12NAL	15NAL	18NAL	21NAL
Lid—3	4NAL	7NAL	26NAL	25NAL	16NAL	19NAL	23NAL
Weld—1	1NA	4NA	7NA	10NA	14NA	17NA	24NA
Weld—2	2NA	5NA	8NA	11NA	15NA	18NA	25NA
Weld—3	3NA	6NA	9NA	12NA	16NA	19NA	26NA
Tube—1	1NA	4NA	7NA	10NA	14NA	17NA	24NA
Tube—2	2NA	5NA	8NA	11NA	15NA	18NA	25NA
Tube—3	3NA	6NA	9NA	12NA	16NA	19NA	26NA

Table A2.
Final set of three material curves for the lid, weld, and tube (Wall) at each temperature.

material strength. Nonetheless, the following is pursued as though the weld material has different stress-strain curve data because this was the original plan in the project and illustrates how three different materials would be handled.

For the wall material, there were three stress-strain (ss) curves at each temperature. For the lid material, there were four to five replicates at some temperatures as shown in **Table A1**. These were reduced to three curves at each temperature by first determining which curve had the lowest effective strength to failure (lowest predicted failure pressure in the PCAP can simulation) and which curve had the highest effective strength to failure. Then, the medium strength curve that resulted in a computed failure pressure closest to the middle between the failure pressures with the high and low strength curves was identified.

Ranking and down-selection were conducted with mechanical-only isothermal simulations with Mesh 4. The effective strength of the material curves was determined according to the calculated pressure at which weld critical TP and EQPS values from **Table 1** were reached. A uniform temperature condition at the relevant temperature from **Table A1** and a linear pressure ramp of 63 psi/min representative of the reference can/test #6 were used in the simulations.

To down-select the lid material curves for temperatures with more than three curves, simulations were conducted for each ss curve at 20, 200, 400, and 800°C. These simulations used wall and weld ss curves labeled Tube-1 and Weld-1 in **Table A1** at the said temperatures.

Table A2 shows the final three material curves chosen for the lid. It should be noted that the order of the lid material curves at each temperature does not necessarily coincide with the high, medium, and low rankings. The final ranking of the lid material curves is reevaluated in the next phase of this process.

The next phase involved running mechanical-only simulations with various combinations of the tube (T), lid (L), and weld (W) ss curves at each temperature. The ranking process included 6 rounds of simulations as exemplified in **Table A3** for the case of 700°C. In the first three rounds, each of the three replicate curves was sampled starting with the weld in Round 1, then the lid in Round 2, and finally the tube in Round 3. An example ranking analysis for Round 1 is explained immediately after **Table A3**. Rounds 4 and 5 rechecked the rankings for both the weld

Round	T	L	W	Comments
1	1	1	1,2,3	Three runs to determine weld curve rankings
2	1	3,2,(1)	2	Two runs to determine lid curve rankings using the medium weld curve (here W = 2). Note that the L = 1 simulation was previously performed in Round 1
3	2,3,(1)	2	2	Two runs to determine tube curve rankings using the medium weld (W = 2) and lid (L = 2) curves. Note that T = 1 simulation was performed in Round 2
4	3	2	1,(2),3	Two runs to **recheck** weld curve rankings from Round 1. Use the medium tube (T = 3) and lid (L = 2) curves. Note that the W = 2 simulation was performed in Round 3
5	3	3,(2),1	2	Two runs to **recheck** lid curve rankings from Round 2. Use the medium tube (T = 3) and weld (W = 2) curves. Note that the L = 2 simulation was performed in Round 4
6	2,3,1	3	1	Three runs to **recheck** tube curve rankings from Round 3. Use the low lid (L = 3) and low weld (W = 1) strength curves

Table A3.
700°C example of process used to rank replicate material curves at a given temperature.

and the lid curves using the nominal curves of the other two materials because the initial tests were not necessarily performed using their medium curves. Finally, Round 6 rechecked the rankings for the tube curves using off-medium conditions, in particular at low-low material curve strengths of the lid and weld, since the original tube rankings were conducted using the medium lid and weld curves.

This process was performed for all seven temperatures. In all rechecked cases, the result of material curve rankings remained the same with the one exception of the tube curves at 20°C, which changed low to high strength ordering from 3-2-1 to 3-1-2.

Six rounds of simulations were involved. Numerical indexes in columns 2–4 are from **Table A2**. Left-to-right order for multiple entries in a cell is lowest to highest effective curve strength. Entries in parenthesis ( ) indicate no new simulation was needed; result already available from a prior round.

Figure A1 shows an example of the computed spatial-maximum tearing parameter (TP) in the weld as a function of time (which is linearly related to pressure for these linear pressure-ramp simulations) for a 700°C temperature in Round 1. In this round, the lid-1 and tube-1 ss curves for 700°C in **Table A2** were used as indicated

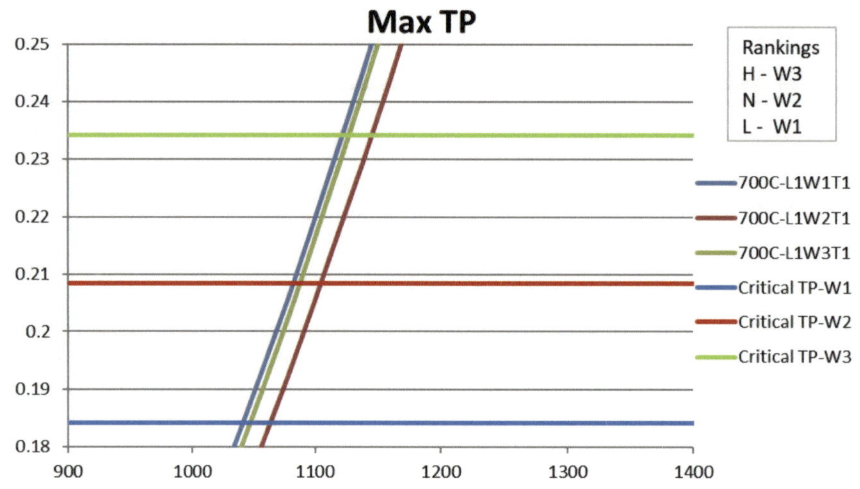

Figure A1.
Comparison of weld spatial-maximum TP results to corresponding critical TP values (plotted horizontal curves) at 700°C.

	20°C	100°C	200°C	400°C	600°C	700°C	800°C
Lid—H	4NAL	7NAL	8NAL	25NAL	15NAL	17NAL	20NAL
Lid—M	3NAL	6NAL	26NAL	11NAL	14NAL	18NAL	23NAL
Lid—L	2NAL	5NAL	9NAL	12NAL	16NAL	19NAL	21NAL
Weld—H	2NA	4NA	8NA	12NA	15NA	19NA	24NA
Weld—M	1NA	5NA	9NA	11NA	14NA	18NA	25NA
Weld—L	3NA	6NA	7NA	10NA	16NA	17NA	26NA
Tube—H	3NA	4NA	9NA	12NA	16NA	17NA	24NA
Tube—M	1NA	5NA	7NA	10NA	14NA	19NA	25NA
Tube—L	2NA	6NA	8NA	11NA	15NA	18NA	26NA

Table A4.
Final material curve rankings for the lid, weld and tube.

in the applicable row of **Table A3**, while the weld material ss curves varied over the three identified for 700°C in **Table A2**. The calculated rises of the material damage TP values in time are compared to their critical TP values from **Table 1** (plotted horizontal lines) to indicate failure by these criteria. The first result (W1) to reach its critical TP value was designated as low strength (L), the second result (W2) was designated as medium (M), and the third result (W3) was designated as high strength (H).

A similar process was used to evaluate the material curve rankings in all rounds of the ranking process. From **Figure A1**, note that at any point in time the 700°C W2 ss curve yields lowest calculated damage (TP value) of any of the weld ss curves, so it represents the "strongest" ss curve by this measure. However, the TP response reaches its critical value faster than the W1 ss curve, which is higher in effective "strength-to-failure." The latter is the measure used for ranking the effective strength of the ss curves.

The final curve strength rankings for the lid, weld, and tube are summarized in **Table A4**. Approximately, 116 mechanical-only simulations were performed in the ranking process, 18 to reduce the number of lid curves to three at 20, 200, 400, and 800°C, and 14 for each of the 7 temperatures to rank the curve strengths. For lid and weld ss curves, the rankings were consistent whether a critical TP or critical EQPS value was used to indicate failure. However, some of the tube results did show differences, and in those cases, the TP ranking was used because the tearing parameter was believed to be the most valid criterion for this application (less mesh-related error than with EQPS; see [3]). Note that even though the same three stress-strain curves per temperature are used for the can weld and walls, the curve strength rankings at a given temperature are usually different for these can parts. This reflects the dependency of effective curve strength on the particular geometry and loading conditions.

Author details

Vicente Romero*, Amalia Black, George Orient and Bonnie Antoun
Sandia National Laboratories, Albuquerque, NM, USA

*Address all correspondence to: vjromer@sandia.gov

References

[1] Black A, Romero V, Breivik N, Orient G, Antoun B, Dodd A, et al. Predictive capability assessment project: Abnormal thermal-mechanical breach V&V/UQ. Sandia National Laboratories report SAND2019-13790 (Official Use Only/Export Controlled). November 2019

[2] Suo-Anttila JM, Dodd AB, Jernigan DA. Thermal mechanical exclusion region barrier breach foam experiments (800C upright and inverted 20 lb/ft3 PMDI Cans). Sandia National Laboratories report SAND2012-7600 (OUO/ECI). September 2012

[3] Black A, Romero V, Breivik N, Orient G, Suo-Anttila J, Antoun B, et al. Verification, validation, and uncertainty quantification of a thermal-mechanical pressurization and breach application. In: Presentation VVS2015-8047 in the Archives of the ASME Verification & Validation Symposium. Las Vegas, NV; May 13-15, 2015

[4] Romero V, Black A, Dodd A, Orient G, Breivik N, Suo-Anttila J, et al. Real-space model validation-UQ methodology and assessment for thermal-chemical-mechanical response and weld failure in heated pressurizing canisters. ASME Journal of Verification, Validation and Uncertainty Quantification. 2019

[5] Romero V, Black A. Processing of random and systematic experimental uncertainties for real-space model validation involving stochastic systems. ASME Journal of Verification, Validation and Uncertainty Quantification.

[6] Romero V. Propagating and combining aleatory uncertainties characterized by continuous random variables and sparse discrete realizations from random functions. Sandia National Laboratories document SAND2019-14642 C, 22nd Non-Deterministic Approaches Conference, AIAA SciTech. Orlando, FL; Jan 6-10, 2020

[7] Romero V, Black A. Adaptive polynomial response surfaces and level-1 probability boxes for propagating and representing aleatory and epistemic components of uncertainty in model validation. Sandia National Laboratories document in review. 2019

[8] Antoun BR. Material Characterization and Coupled Thermal-Mechanical Experiments for Pressurized, High Temperature Systems. Technical Report. Livermore, CA: Sandia National Laboratories; 2012

[9] Frost HJ, Ashby MF. Deformation-Mechanism Maps: The Plasticity and Creep of Metals and Ceramics. Oxford [Oxfordshire]: Pergamon Press; 1982

[10] Lichtenfeld JA, Mataya MC, Van Tyne CJ. Effect of strain rate on stress-strain behavior of alloy 309 and 304L austenitic stainless steel. Metallurgical and Materials Transactions A. 2006; **37A**:147-161

[11] Wellman GW. A simple approach to modeling ductile failure. Sandia National Laboratories report SAND 2012-1343. June 2012

[12] Sierra/SM Development Team. Sierra/SM theory manual. Sandia National Laboratories report SAND2013-4615. July 2013

[13] Chen W-F, Han DJ. Plasticity for Structural Engineers. Ft. Lauderdale, FL: J. Ross Publishing; 2007

[14] Available from: http://www.mathworks.com/matlabcentral/fileexchange/13812-splinefit

[15] Wilson KM, Karlson KN, Jones R, Hoffa T. MatCal: A tool for improving

the traceability and workflow for material calibration. Sandia National Laboratories document SAND2019-11952 C (Official Use Only/Export Controlled). October 2019

[16] Adams BM, Bauman LE, Bohnhoff WJ, Dalbey KR, Ebeida MS, Eddy JP, et al. Dakota, a multilevel parallel object-oriented framework for design optimization, parameter estimation, uncertainty quantification, and sensitivity analysis: Version 6.0 user's manual. Sandia Technical Report SAND2014-4633. July 2014

[17] Brozzo P, Deluca B, Rendina R. A new method for the prediction of the formability limits of metal sheets. In: Proceedings of the 7th Biennial Congress of International Deep Drawing Research Group. 1972

[18] Notz PK, Subia SR, Hopkins MM, Moffat HK, Nobel DR. Aria 1.5: User manual. Sandia National Laboratories report SAND2007-2734. April 2007

[19] Erickson KL, Dodd AB, Hogan RE. Modeling pressurization caused by thermal decomposition of highly charring foam in sealed containers. In: Proceedings of BCC 2010, Stamford, CT, 23-26 May 2010

[20] Edwards HC, Stewart JR. SIERRA: A software environment for developing complex multi-physics applications. In: Bathe KJ, editor. First MIT Conference on Computational Fluid and Solid Mechanics. Amsterdam: Elsevier; 2001. pp. 1147-1150

[21] Larsen ME, Dodd AB. Modeling and validation of the thermal response of TDI encapsulating foam as a function of initial density. Sandia National Laboratories report SAND2014-17850. September 2014

[22] Contact: Nicole Breivik, Sandia National Laboratories, Laser weld modeling methods for deformation and failure

[23] Romero V, Dempsey F, Antoun B. Application of UQ and V&V to experiments and simulations of heated pipes pressurized to failure. In: Mehta U, Eklund D, Romero V, Pearce J, Keim N, editors. Chapter 11 of Joint Army/Navy/NASA/Air Force (JANNAF) e-book: Simulation Credibility—Advances in Verification, Validation, and Uncertainty Quantification, Document NASA/TP-2016-219422 and JANNAF/GL-2016-0001. 2016

[24] Romero V, Dempsey JF, Wellman G, Antoun B. A Method for projecting uncertainty from sparse samples of discrete random functions — Example of multiple stress-strain curves. Paper AIAA-2012-1365, 14th AIAA Non-Deterministic Approaches Conference; April 23-26, 2012; Honolulu, HI

[25] Romero V, Dempsey JF, Schroeder B, Lewis J, Breivik N, Orient G, et al. Evaluation of a simple UQ approach to compensate for sparse stress-strain curve data in solid mechanics applications. In: 19th AIAA Non-Deterministic Approaches Conference, Paper AIAA2017-0818, AIAA SciTech 2017, Jan. 9-13, Grapevine, TX

[26] Romero V, Bonney M, Schroeder B, Weirs VG. Evaluation of a class of simple and effective uncertainty methods for sparse samples of random variables and functions. Sandia National Laboratories report SAND2017-12349. November 2017

[27] Romero V, Schroeder B, Dempsey JF, Breivik N, Orient G, Antoun B, et al. Simple effective conservative treatment of uncertainty from sparse samples of random variables and functions. ASCE-ASME Journal of Uncertainty and Risk in Engineering Systems: Part B. Mechanical Engineering. 2018;**4**: 041006-1-041006-17. DOI: 10.1115/1.4039558

[28] Jamison R, Romero V, Stavig M, Buchheit T, Newton C. Experimental data uncertainty, calibration, and validation of a viscoelastic potential energy clock model for inorganic sealing glasses. In: Sandia National Laboratories Document SAND2016-4635C, Albuquerque, NM, Presented at ASME Verification & Validation Symposium; Las Vegas, NV; May 18-20, 2016

[29] Romero V, Heaphy R, Rutherford B, Lewis JR. Uncertainty quantification and model validation for III-V SSICs in annular core research reactor shots. Sandia National Laboratories report SAND2016-11772 (Official Use Only/ Export Controlled). 2016

[30] Romero V. Real-space model validation and predictor-corrector extrapolation applied to the sandia cantilever beam end-to-end UQ problem. In: Paper AIAA-2019-1488, 21st AIAA Non-Deterministic Approaches Conference, AIAA SciTech 2019; Jan. 7-11; San Diego, CA

[31] Romero V, Mullins J, Swiler L, Urbina A. A comparison of methods for representing and aggregating experimental uncertainties involving sparse data—more results. SAE International Journal of Materials and Manufacturing. 2013;**6**(3):447-473. DOI: 10.4271/2013-01-0946

[32] Romero V, Swiler L, Urbina A, Mullins J. A comparison of methods for representing sparsely sampled random quantities. Sandia National Laboratories report SAND2013-4561. September 2013

[33] Bhachu KS, Haftka RT, Kim NH. Comparison of methods for calculating B-basis crack growth life using limited tests. AIAA Journal. 2016;**54**(4): 1287-1298

[34] Zaman K, McDonald M, Rangavajhala S, Mahadevan S. Representation and propagation of both probabilistic and interval uncertainty. In: Paper AIAA-2010-2853, 12th AIAA Non-Deterministic Approaches Conference; April 12–15, 2010; Orlando, FL

[35] Pradlwarter HJ, Schuëller GI. The use of kernel densities and confidence intervals to cope with insufficient data in validation experiments. Computer Methods in Applied Mechanics and Engineering. 2008;**197**(29-32): 2550-2560

[36] Sankararaman S, Mahadevan S. Likelihood-based representation of epistemic uncertainty due to sparse point data and/or interval data. Reliability Engineering and System Safety. 2011;**96**(7):814-824

[37] Sankararaman S, Mahadevan S. Distribution type uncertainty due to sparse and imprecise data. Mechanical Systems and Signal Processing. 2013; **37**(1):182-198

[38] Hahn GJ, Meeker WQ. Statistical Intervals—A Guide for Practitioners. New York: Wiley & Sons; 1991

[39] Montgomery DC, Runger GC. Applied Statistics and Probability for Engineers. New York: Wiley & Sons; 1994

[40] Howe WG. Two-sided tolerance limits for normal populations—Some improvements. Journal of the American Statistical Association. 1969;**64**:610-620

[41] Young DS. Tolerance: An R package for estimating tolerance intervals. Journal of Statistical Software. 2010; **36**(50):1-39

[42] Winokur J, Romero V. Optimal design of computer experiments for uncertainty quantification with sparse discrete sampling. Sandia National Laboratories document SAND2016-12608. 2016

[43] Romero V, Winokur J, Orient G, Dempsey JF. Confirmation of discrete-direct calibration and uncertainty propagation approach for multi-parameter plasticity model calibrated to sparse random field data. In: Presentation VVS2019-5172 in the Archives of the ASME Verification & Validation Symposium; May 15-17, 2019

[44] Conover WJ. On a better method for selecting values of input variables for computer codes. Unpublished 1975 manuscript recorded as Appendix A of "Latin Hypercube Sampling and the Propagation of Uncertainty Analysis of Complex Systems," Helton JC and Davis FJ, Sandia National Laboratories report SAND2001-0417 November 2002

[45] McKay MD, Beckman RJ, Conover WJ. A comparison of three methods for selecting values of input variables in the analysis of output from a computer code. Technometrics. 1979; **21**(2):239-245

[46] Helton JC, Davis FJ. Latin hypercube sampling and the propagation of uncertainty in analyses of complex systems. Reliability Engineering and System Safety. 2003; **81**(1):23-69

[47] Jekel C, Romero V. Bootstrapping and jackknife resampling to improve sparse-data UQ methods for tail probability estimates with limited samples. In: ASME Paper VVS2019-5127, ASME 2019 Verification and Validation Symposium VVS2019; May 15-17, 2019; Las Vegas, NV

[48] Jekel C, Romero V. Improving tail probability estimation from sparse-sample UQ methods with bootstrapping and jackknifing. ASME Journal Verification, Validation and Uncertainty Quantification. Sandia National Laboratories document SAND2019-10731 J

[49] Jekel C, Romero V. Conservative estimation of tail probabilities from limited sample data. Sandia National Laboratories Report in Review. 2019

Chapter 3

Stress Corrosion Cracking Behavior of Materials

Alireza Khalifeh

Abstract

Stress corrosion cracking is a phenomenon associated with a combination of tensile stress, corrosive environment and, in some cases, a metallurgical condition that causes the component to premature failures. The fractures are often sudden and catastrophic, which may occur after a short period of design life and a stress level much lower than the yield stress. It can also occur after several years of satisfactory services due to operating errors and changing process conditions. Two classic cases of stress corrosion cracking are seasonal cracking of brass in ammoniacal environment and sensitization and stress corrosion cracking of stainless steels in existence of chlorides, caustic, and polythionic acid. Presence of crack and other defects on the material surfaces accelerates the fracture processes. Therefore, when designing components, the role of imperfections and aggressive agents together must be taken into account. The fracture mechanic introduces a material characteristic namely fracture toughness or $K_{ISCC} = \sigma\sqrt{\pi a}$, which properly describes the fracture behavior of materials in such conditions. The main objective in writing of this chapter is to present scientific findings and relevant engineering practice involving this phenomenon.

Keywords: stress corrosion cracking, fracture mechanic, mechanism

1. Introduction

Stress corrosion cracking (SCC) is service failure of engineering materials that occurs by slow, environmentally induced crack propagation. Identification of SCC occurred between 1930s and 1950s, the mechanism of SCC was explained between 1960s and 1970s and its application and development began at 1980s [1]. In fact, the industry owners did not find the austerity of the problems connected to the stress corrosion cracking until the tragic rupture of a digester at Pine Hill, Alabama, in 1980 [2]. After the accident, considerable investigations have been taken on SCC phenomenon. In this type of failure, the crack commencement and growth is the consequence of reciprocal action of sensitizes material, tensile stress, and corrosive environment.

Various types of SCC have been distinguished. Chloride SCC appears in austenitic stainless steels under tensile mechanical stress in the existence of chloride ion, oxygen, and a high temperature condition [3]. Cracking of stainless steels under caustic environments in the presence of a high hydrogen concentration is known as caustic embrittlement [4]. SCC cracking of steels in hydrogen sulfide environment is confronted in oil industries [3]. Cracking of brass in ammonia environments is another type of SCC, which is known as sessional cracking.

During SCC phenomenon, the material is essentially unattacked over most of its surface, while fine and branch cracks develop into the bulk of material [5]. This cracking phenomenon has serious consequences since it can occur under stress levels much lower than the designer intended and cause the equipment and structural elements to catastrophic fractures [6–10].

In this chapter, the feature of SCC is first introduced. Then, the requirements of the stress corrosion cracking are expressed. In next step, the mechanisms attributed to this corrosion phenomenon are discussed in detail. Designing of structural elements and mechanical equipment based on the tensile properties may be led to false results. Accordingly, the science of fracture mechanics applies in situation where SCC can be occurred. Fracture mechanic presented criteria that considered the role of imperfection on the fracture of materials. Next, based on the corrosion science and empirical data, the methods of preventions are presented. Finally, case studies are introduced to better understand of the issue.

2. SCC features

In macroscopic scale, SCC failures appear to be brittle even if the material is ductile, and stress level is lower than design stress. The cracks in SCC are not mechanical but are caused by corrosion. A typical growth and developments of cracking by cause of the stress corrosion cracking phenomenon in AISI316L stainless steel are shown in **Figure 1a**.

The branches developed, intergranular and transgranular cracks are the character of stress corrosion cracking in microscopic levels, **Figure 1b**. In intergranular mode, the cracks proceed along grain boundaries while transgranular cracks grow across the grains [3]. The mode of crack proceeds depends on the material microstructure and environment.

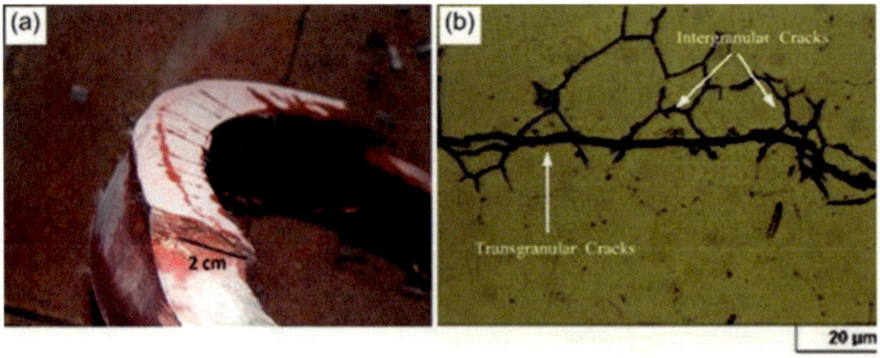

Figure 1.
Macroscopic and microscopic feature of the stress corrosion cracking: (a) Cross-section of a failed tube and formation of macro branching cracks on it. (b) Optical micrograph of a typical SCC in AISI316L stainless steel exposed to chlorides [3, 4].

3. Requirement for SCC

Three key elements are essential for initiation and growth of the stress corrosion cracking: a sensitized material, a specific environment, and adequate tensile stress. It is shown in **Figure 2**.

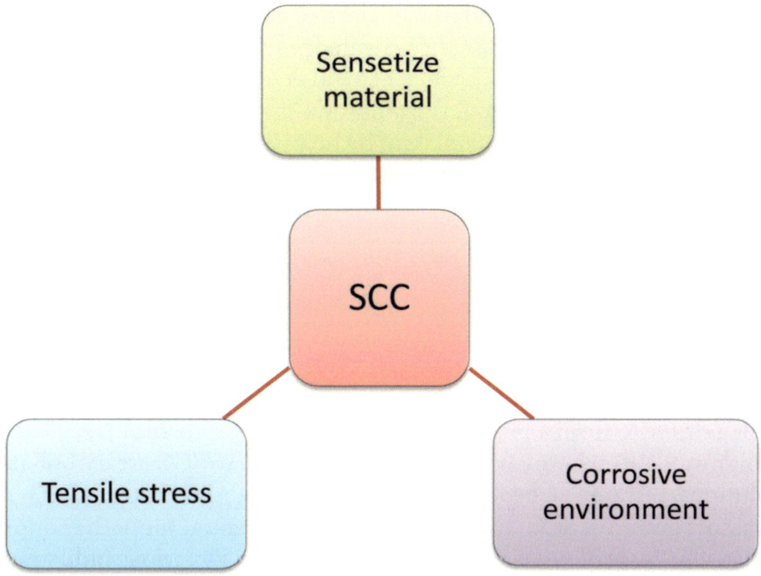

Figure 2.
Essential parameters for occurring SCC.

3.1 Materials

3.1.1 Stainless steels

When austenitic stainless steel grades contain more than 0.03 wt% that are exposed to temperature from 415 to 850°C, their microstructure becomes sentient to precipitation of chromium carbides ($M_{23}C$) along grain boundaries called as sensitization [11, 12]. Constitution of Cr enrichment carbides along with grain boundaries may severely evacuate the area aside to the grain boundaries from free chromium and render them sensitive to rapid corrosion in presence of aggressive agents such as chloride, caustic, and polythionic acid [4, 12, 13]. Grain boundary dissolution aside tensile tress lead the component made of these ductile alloys to brittle fracture. An example of the SCC failure in an austenitic stainless steel is shown in **Figure 1**.

3.1.2 Copper and copper alloys

Copper is alloyed with zinc, tin, nickel, aluminum, and silicon to produce various types of brass, bronze, cupronickel, aluminum bronze, and silicon bronze, respectively. Most of SCC failure mechanisms for copper-based alloys have been predominantly investigated for ammonia-brass systems [14]. A classic example of SCC for brass systems is season cracking. It specifies the stress corrosion cracking of brass cartridge cases of British forces in India [11]. Two mechanism have been found to be attributed to SCC of copper-based alloys: (a) the passive film rupture and (b) de-alloying [15]. According to first theory, intergranular SCC is to occur due to passive-film rupture and transient dissolution mechanism in copper-zinc alloys. Chen et al. reported that aluminum bronze (Al-Cu-Zn) underwent SCC in fluoride-containing solutions through the film-rupture mechanism [16]. According to de-alloying, that is, selective segregation of the alloying elements, for example, Zn in Cu-Zn systems, is the primary factor in the stress corrosion cracking phenomenon. For example, in the case of Cu-Zn alloy systems, preferential segregation of Zn take place in the crack tip region, or along the grain boundaries. Domiaty and Alhajji found that cupronickel alloy Cu-Ni (90–10) underwent SCC in seawater polluted with sulfide ions due to selective dissolution of copper [17]. Chen et al. also reported that de-alloying resulted in SCC of aluminum brass (Al-Cu-Zn) in fluoride-containing aqueous solutions [16]. The Pourbaix diagrams for pure metals can be helpful in anticipating the sensitivity to de-alloying and in appraising the tendency of selective dissolution and hence the SCC behavior [16].

3.1.3 Carbon steels

Stress corrosion cracking of carbon steels has been identified as one of the main reasons leading to leakage and rupture events of pipe lines and steam generator boilers with catastrophic consequences [18, 19]. Carbon steels have shown SCC in conditions that there is a desire to form a protective passive layer or oxide film [13, 20–22]. In fact, the environments where carbon steel is susceptible to SCC are carbonates, strong caustic solutions, nitrates, phosphates, and high-temperature water [13, 23–25]. The problems associated with SCC in carbon steels are influential for both economic and safety reasons, because of widespread use of these alloys in various industries.

3.1.4 Aluminum and aluminum alloys

Aluminum and its alloys are extensively used in military, aerospace, and structural applications. These groups of engineering material under specific environment and sufficient magnitude of stress are susceptible to SCC [26, 27]. Determinant factor on the stress corrosion cracking behavior in Al alloy systems is the chemical composition. Alloying elements have strong influences on the formation and stability of the protecting layer on the surface of alloy and also may show effects on the strength, grain size, grain boundary precipitations, and magnitude of residual stress within the material [28, 29]. In between them, series of 2xxx, 7xxx, and 5xxx (alloys containing magnesium) are sensitive to SCC [28]. Failures associated with SCC in aircraft constituents made of 7075-T6, 7079-T6, and 2024-T3 Al-alloys were reported between 1960 and 1970 [27].

Another important factor on SCC behavior of aluminum alloys is type of heat treatment and quenching rates [26, 30]. Heat treatment shows a crucial role in formation and organization of the constituent particles in order to attain a high strength aluminum alloys. It has been reported that aluminum alloy series 7xxx are sensitive to cooling rate and many researchers investigated the effect of cooling rates on the stress corrosion cracking behavior of the alloys. Researches have displayed that a decline in the cooling rate results in a rise in the size and inter particle distance of the GBP forward with an increase in width of the PFZ. Unfortunately, this leads to a decrease of copper contents in the grain boundary of precipitates and reduction of resistance to SCC [28–30].

There is a significant discrepancy in the texts on the exact mechanism of stress corrosion cracking in Al alloys. Evidence shows that the proposed mechanism of SCC pertains on the alloy system in question: anodic dissolution is suggested for the 2xxx alloy systems (Al-Cu and Al-Cu-Li), while hydrogen-induced cracking or HIC is presented by most investigators in the 5xxx and 7xxx alloys. Another mechanism may be properly explained SCC in some Al alloy systems is the rupture passive film. Thus, while SCC phenomenon of Al alloys is well documented, the actual mechanisms are still challenged [29].

Two methods are suggested improving SCC resistance in Al alloys. The first method is to break down the continuity of the GBPs and the second is to decline the galvanic potential nonconformity between the grain boundaries and matrix.

3.1.5 Titanium and titanium alloys

Titanium alloys are broadly utilized in aerospace industry because of their high corrosion resistance and specific strength. Two widely used grades of Ti are Ti-6Al-4 V known as Ti-64 alloy and Ti-8Al-1Mo1V known as Ti-811. The latter is specified by its lower density and surpassing stiffness, but unfortunately is susceptible to SCC.

The significant corrosion resistance of Ti alloys under oxidizing media is a result of formation of the protective passive TiO_2 film on the surface. For Ti-811 grade, instinctive passivation arises in 3.5% sodium chloride HCl solution [31]. However, the passive protective layer will be destroyed if load applied is over the yield stress of the oxide film, and corrosive environment will then expose to the fresh metal and lead to SCC [32]. In Ti alloys systems, oxygen and aluminum additions enhance SCC sensitivity and change the slip condition to a planar slip [33]. Slip plan growth of SCC cracks in this grade of Ti alloys is shown in **Figure 3**.

3.1.6 Polymers

SCC can be occurred in polymers, when the components made of these materials are exposed to specific aggressive agents like acids and alkalis. Similar to metals combination of polymer and environment which lead to SCC is specific. For instance, polycarbonate is susceptible to SCC in alkalis, but not by acids. On the other hand, polyesters are prone to SCC when exposed to acids [34]. Another form of SCC in polymeric materials is ozone attack [35]. Small traces of the gas in the air will attack polymer bonds and lead them to degradation. Natural rubber, nitrile butadiene rubber, and styrene-butadiene rubber were found to be most sensitive to this form of failure. Ozone cracks are very dangerous in fuel pipes because the cracks can grow from external surfaces into the bore of the pipe, and so fuel leakage and fire will be its consequences [36–38]. Investigations have shown a critical amount of deformation and ozone concentration require for SCC crack growth in rubbers [35].

3.1.7 Ceramics

The stress corrosion cracking is less common in ceramic materials that are more resistant to chemical attacks. However, phase transformation under stress is usual in the ceramic substances such as zirconium dioxide. This phenomenon led them to toughening [39]. Silica or SiO_2 is extensively used in microelectromechanical

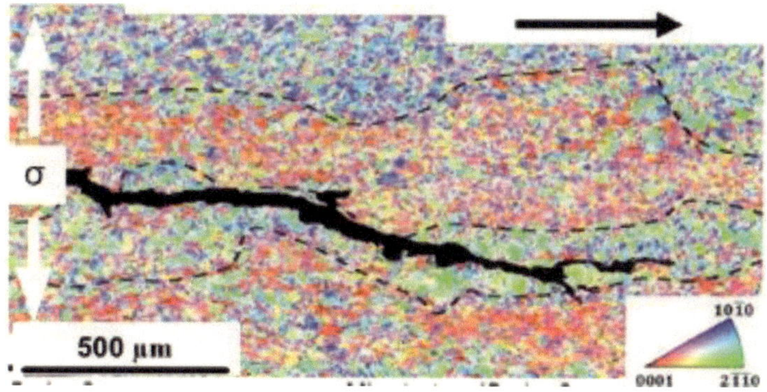

Figure 3.
IPF map of the SCC crack growth in a wrought Ti-811 specimen. Crack propagation extended along planar slip (0001) [33].

systems and in integrated circuits as sacrificial release layers and dielectric layers. Silica is sensitive to SCC in air because of the reaction between Si-O bonds with H_2O molecules under stress [40].

3.2 Environments

The environment prone to stress corrosion cracking is special since not all environments provide condition for cracking. The corrosive condition may be a permanent service environment, such as processes water for a heat exchanger, or temporary environments produced by an operational errors, like deposits form on the equipment during shut downs in chemical plants. Moreover, environments that lead the components to SCC are often aqueous in nature and can be either a thin layer of condensed moisture or a volume solution [41]. Generally, the stress corrosion cracking occurs in aqueous medium, but it also occurs in certain metals, fused salts, and inorganic liquids [11]. The aggressive environment to passive film of stainless steels is chlorides, caustic, and polythionic acid. The austenitic stainless steel series 300 like AISI304 and AISI316 are more sensitive in environment consisting of chlorides. It was reported that the SCC in austenitic stainless steels containing chlorides proceeds transgranulary and often develops at a temperature over 70°C [42, 43]. Caustic embrittlement or the SCC induced by caustic is another severe problem in the equipment made of austenitic stainless steels and led to many explosions in the steam boilers and super heaters [4, 44, 45]. It was also seen that the existence of sulfur in feed gas in petrochemical plants causes forming of polythionic acid (H_2SxO, x = 2 to 5), which in presence of moisture also induced intergranular SCC in austenitic stainless steels [46, 47].

Carbon steels are found to be sensitive to SCC in nitrate, NaOH, acidic H_2S, and seawater solutions [48]. Soil environment is responsible for SCC cracking of underground pipeline carbon steels where cathodic protection is utilized to preserve lines against general corrosion [49, 50]. Cathodic protection of buried pipelines can result in the production of alkali solutions at the pipe surface where the coating is disbanded. The formed alkalis consist of sodium hydroxide, carbonate, and bicarbonate solution. Investigations have shown that in the presence of these compounds and some range of potential protection carbon steel pipes fail due to the stress corrosion cracking. For instance, carbonate solution and a corrosion potential [between −0.31 and −0.46 V (SHE)] developed the stress corrosion cracking in a high strength carbon steel pipeline [49, 50].

Environment conditions that lead the copper alloys to SCC are ammonia, amines, and water vapor [48]. The source of ammonia may be from decomposition of organic materials [11]. In ammonia production plans, ammonia comes from leakage of the equipment and piping systems, when these components are not completely leak-proof.

In aluminum and aluminum alloy, pure aluminum is safe to SCC. On the other hand, duralumin alloy refers to aluminum-copper alloys, under tension stress aside of moisture, and it may fail due to cracking along the grain boundaries. Aluminum alloys are also susceptible in SCC in NaCl solution [50].

The stress corrosion cracking of pure titanium was observed by Kiefer and Harple in red fuming nitric acid for the first time [51]. Titanium alloys are also susceptible to SCC when is exposed to sea water and methanol-HCl [52].

3.3 Stress

The stresses that cause SCC are directly applied stress, thermal operational stress, residual stress or combination of all [3]. The operational stress consists of operational applied stress and thermal stress. The operational applied stress is considered by designers and usually calculated less than the yield strength. The thermal stress is generated due to thermal cooling and heating of component during services and shut downs. The thermal stress is calculated from thermal strains that are expressed by a temperature-dependent differential expansion coefficient [13]:

$$\{\Delta\varepsilon^{th}\} = [\alpha] \cdot \Delta T \tag{1}$$

in which, $\Delta\varepsilon^{th}$ is the variation of strain, α is the thermal expansion coefficient of material, and ΔT is the changing in temperature.

The main source of stress attributed to stress corrosion cracking comes from the residual stress. The well-known sources of the residual stress are welding and fabrication processes. Residual stress due to welding plays an imperative role in the stress corrosion cracking of metal alloys. In this area, two factors are determined. First, the welding residual stress by the cause of nonuniform temperature changes during welding, which can be computed by Eq. (1). Second, in some of the steel

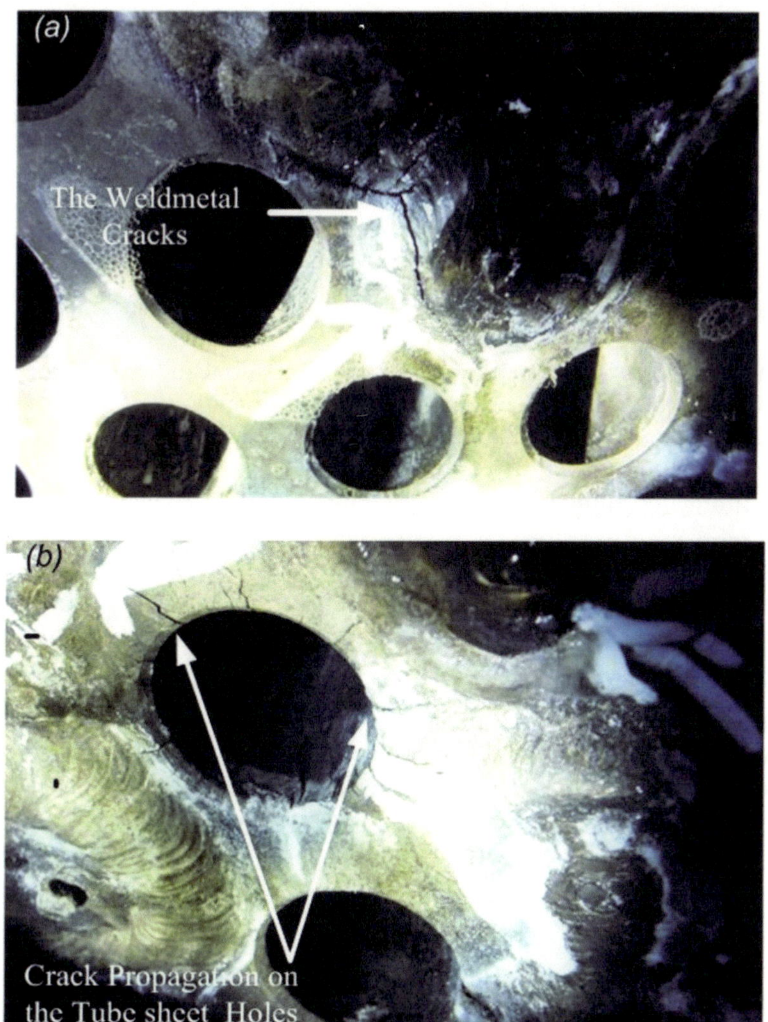

Figure 4.
Two sources of stress in the SCC: (a) welding residual stress and (b) mechanical residual stress [3].

grades, the solid-state austenite to martensite transformation in the time of cooling generates a significant value of the residual stresses [53]. In carbon and low alloy steels, martensite is formed by quenching of the austenite containing carbon atoms at such a fast cooling rate that carbon atoms do not have a chance to diffuse out from the crystal structure and form cementite. The trapped carbon atoms in octahedral interstitial sites of iron atoms, generating a body-centered tetragonal (bct) structure that is in super saturation, relate to the ferrite with body-centered cubic (bcc) structure [54]. Therefore, during this phase transformation, the volume of metal increases, and a significant residual stress is also produced [54]. In austenitic stainless steel, there are no phase transformation processes, and so this mechanism does not attribute in their welding residual stress. An example of SCC in the weld area of an austenitic stainless steel component is shown in **Figure 4a**.

The product manufacturing systems that generate considerable residual stresses include casting, rolling, forging, drilling, machining, heat treatments, cooling, carburization, and straightening [55]. In many of the material manufacturing processing, there is a rebalancing of the generated residual stresses during or after the production process, and the extent of the final residual stresses is often less than half of the yield stress or σ_y. Some material processing, such as heat treatment with slow cooling rate, may release the stresses generated by previous treatments. But processes like rapid quenching or machining, which create localized yielding at the surface, may leave residual stress in magnitude of the yield stresses at surface or down surface of the material [55]. A typical SCC in heavy machining area of an austenitic stainless steel component is presented in **Figure 4b**.

Finally, the stress must be tensile form, and comprehensive stresses do not cause the equipment to failure due to SCC.

4. Stress corrosion cracking mechanisms

The mechanism of SCC depends on the type of material and environment. Many models have been presented describing SCC phenomenon. Each of these models has its own restraints in that it can be utilized to describe SCC in limited number of metal-environment systems. In fact, there is no unified mechanism attributed in SCC for all metal environment combination. A few presented models are given in following.

4.1 Mechano-electrochemical model

According to this model, there are pre-existing regions in an alloy microstructure that become sensitive to anodic dissolution. For instance, a precipitate in grain boundary may be anodic with respect to the grain boundary and provide an active path for localized corrosion. In the same way, if a nobler phase is precipitated at grain boundaries, adjacent to precipitates provide an active track for localized corrosion. A classic example of this mechanism can be seen in austenitic stainless in which precipitates of chromium carbides along grain boundaries depletes adjacent areas from chromium and provides a path for localized corrosion to be occurred [1].

4.2 Film rupture model

The film rupture mechanism implies for the alloys which passive layer is formed on their surface. In this mechanism, plastic deformation plays the main role. The plastic strain causes film disruption on the surface of metal. After the film disruption, the bare metal is disclosed to environment, and a localized attack at the area of disruption occurs. The processes of disruption strain and film formation are repeated, and cracking growth continues. This mechanism was originally considered for caustic cracking of boilers. It has also been utilized for description of SCC of alpha brass in ammonia environments [1].

4.3 Adsorption phenomenon

The adsorbing model is based on the process of embrittles of the material in the vicinity of a corroding area. First, the model was used to study on failure of high strength martensitic stainless steels in chloride media. According to this model, the adsorption of environmental agents drops the interatomic bond strength and the stress need for a brittle fracture. The fracture mechanic theory properly explains the issue. According to this theory, the theoretical fracture stress required to take apart two layers of inter atomics spacing b is given by:

$$\sigma_{Fr} = \left(\frac{E\gamma}{b}\right)^{1/2} \tag{2}$$

in which, E, γ, and b are Young's modulus, surface energy, and spacing between atom layers, respectively. According to this model, in corroded environments, the aggressive agents are adsorbed at the crack tip, surface energy effectively decreased, and fracture take place in a stress level much lower than the normal condition [1].

5. Application of fracture mechanic

Investigation has shown that the maximum operational stresses and the maximum defect sizes imparted during production processing could limit the life time of materials [56]. Stress produced due applied, thermal or residual, usually below the yield strength of a material. However, surface defects that either preexist or are produced during services by cause of corrosion, wear, or other processes (**Figure 5**) may provide stress concentration and so condition for fracture [41]. Hence, the role of imperfections should take into account. Fracture mechanic introduced another

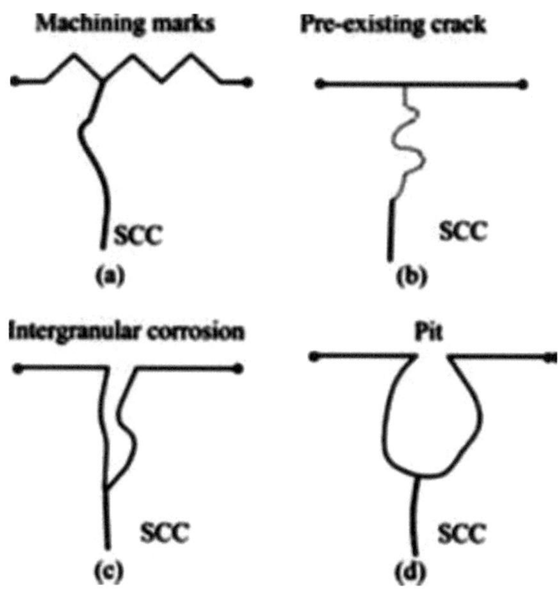

Figure 5.
The sources of SCC sites (a) machining marks, (b) pre-existing crack, (c) intergranular corrosion, and (d) pit [28].

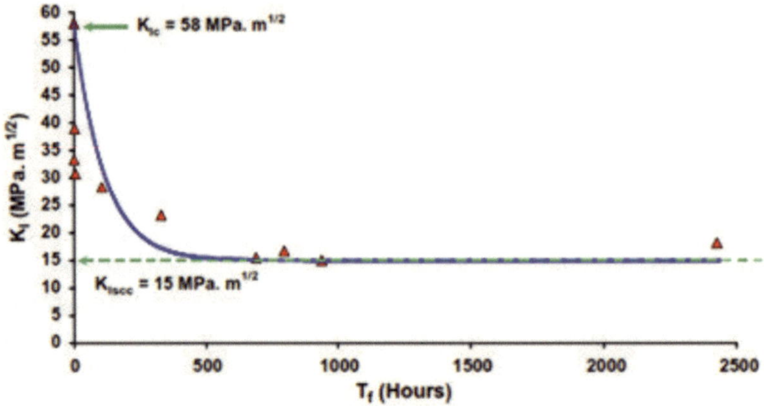

Figure 6.
K_I vs. T_f of AISI4340 steel in 3.5% NaCl at ambient temperature [59].

material property named fracture toughness or K_{Ic} in the same sense that the yield strength. The fracture toughness is a material property that expresses the resistance to crack growth. The relation between fracture toughness, flaw size, and stress in simplest form is expressed as [57]:

$$K_{Ic} = \sigma\sqrt{\pi a} \tag{3}$$

in which, σ denotes stress and a denotes crack length.

According to this relation, there is a critical size of flaw which led a component to brittle fracture. Investigation has shown that for the structure expose to corrosive environments, the scenario is quite different [58]. The corrosive environments cause a significant drop in this property of the materials [59]. This typically is shown in **Figure 6**. As it is seen from the graph, the K_{Ic} in presence of aggressive agents shown by K_{Iscc} is much lower than its value in normal conditions. This means in corrosive environments prone to SCC, a smaller flaw size led the components to catastrophic brittle fractures. Susceptibility of existing cracks to environmentally assisted propagation is a critical consideration for designers while selecting proper materials for environments prone to this form of corrosion [58].

Example 1. Compare the critical crack length for 4340 steel exposed to dry environment and 3.5% NaCl at room temperature:

If we assume that the applied load is on the magnitude of yield stress (Y.S), the critical crack depth, a_{cr}, exists above a size that the stress intensity factor exceeds K_{Ic} or K_{Iscc}. In presence of this condition, the failure occurs in a brittle mode. Based on Eq. (3):

In dry condition: $a_{cr} = \left(\frac{K_{Ic}}{Y.S}\right)^2 = \left(\frac{58}{68}\right)^2 = 0.73$ in.

In 3.5% NaCl condition: $a_{cr} = \left(\frac{15}{68}\right)^2 = 0.05$ in

The results show that the a_{cr} in presence of corrosive agents is about 15 times lower than the normal condition.

6. Prevention

Since the mechanism of SCC is not fully understood, methods of preventing are based on empirical experiences. One or more application of the following methods can be worked out in reduction or prevention of the SCC [11]:

a. Lowering the stress levels. This may be done by annealing treatment in the case of residual stress. Stress relief temperature for plain carbon steels are at a range of 1100–1200°F and for austenitic stainless steels are frequently at temperatures ranging from 1500 to 1700°F.

b. Eliminating aggressive species in the environment. For instance, in case of austenitic stainless steels, reduction of chloride under 10 ppm reduces significantly the probability of the stress corrosion cracking to occur [44].

c. Changing the material in one plane if neither environment condition nor stress level can be changed. For example, in case of AISI304, stainless steel utilizing a high nickel alloy content (e.g., Inconel) can be useful.

d. Applying cathodic protection by impressed current or sacrificial anode. In cathodic protection, note that the failure is due to SCC, not hydrogen embrittlement. In cathodic protection, the released hydrogen due to cathodic reaction can accelerate the hydrogen embrittlement.

e. Adding inhibitors such as phosphate and other organic and inorganic to the system can reduce the stress corrosion cracking effects in mildly corrosive media.

f. Coating is sometimes effective.

g. Shot-peening generates residual compressive stress in surface of the component and prevent the stress corrosion cracking to be occurred.

h. The problem associated with ozone stress cracking can be prevented utilizing anti-ozonants to the rubber before vulcanization treatment.

7. Case studies

7.1 Case 1: stress corrosion cracking failure of a transmission oil product pipeline

In April 2004, an oil transportation pipeline in north part of Iran failed and caused oil leaking. Visual inspections revealed macrocracks and shallow pits on the external surface of the corroded area (**Figure 7**). Optical observation

Figure 7.
Macroscopic observation of the cracking area.

indicated that cracks were started from the bottom of the pits and extended into the wall thickness as shown in **Figure 8**. There was a main singular crack and some branched cracks at the end. The crack features are characteristics of a stress corrosion cracking [60].

7.1.1 Material, environment, and source of stress

The oil transportation pipe line is made of API 5 L X52 carbon steel that is coated by polyethylene tape coating. The utilized polyethylene coating on the external surface of the carbon steel pipe gets loose and separated in some areas. As a consequence, the surface of the underground steel pipeline was uncovered to the wet soil environment. Because of the chemical reactions and the formation of carbonate-bicarbonate solution and arise of underweight stress due to rain fall and land sliding, condition was provided for SCC. Further to SCC, sulfate reducing bacteria (SRB) activities around the pipelines have accelerated corrosion and the failure process.

7.1.2 Mechanism

Two types of SCC on the external surface of underground pipelines have been diagnosticated: classical or high-pH SCC and near neutral or low-pH SCC. In high-pH SCC case, the external cracks more often initiate and progress intergranularly, and in low-pH SCC, cracks extend transgranularly. In this case, as a result of the formation of a high pH carbonate-bicarbonate solution, the cracks have been developed intergranularly (**Figure 7**). By propagation and so increasing the crack depth, chemical composition in the crack's tip changed toward low pH, and the cracks propagated transgranularly (**Figure 7**). In a critical length of crack and the presence of inclusions such as MnS, atomic hydrogen will be formed in crack's tip and penetrates into the microstructure. Formation of hydrogen was due to application of negative cathodic potential, more than 1.2 V with respect to $Cu/CuSO_4$ reference electrode. Investigations have been shown that atomic hydrogen commonly tends to concentrate in the stress concentration points, microstructure interfaces like inclusion and matrix, microvoids, and other defects. The increase in the amount of atomic hydrogen in these sites and combination together decreased the interface cohesively strength and make atomic disbanding. These led the crack growth and fracture toward a cleavage and brittle mode.

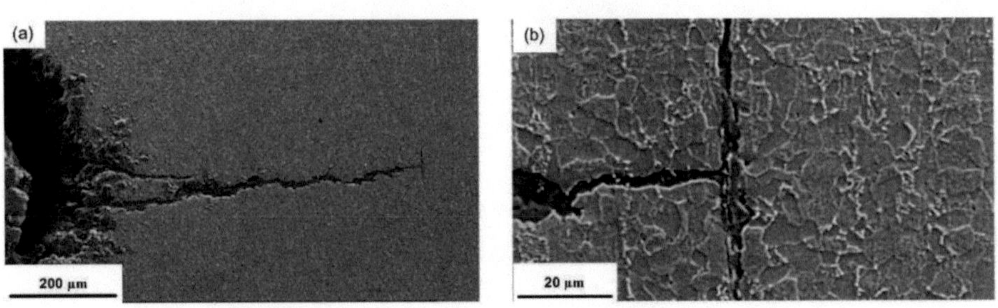

Figure 8.
(a) Initiating of cracks from pitting on the surface of the carbon steel pipe and its branched development (unetched), (b) intergranular crack growth in the microstructure (etched) [60].

7.1.3 Prevention

Utilizing a high level of negative potential (1.2 V Cu/$CuSO_4$) in cathodic protection provides condition for production of carbonate and bicarbonate which induced SCC. Therefore, controlling of the potential can be effective in reduction of these failures.

7.2 Case 2: failure of a hydroprocessing reactor

A vertical hydro-processing reactor failed only after 1 month operation. Penetrant testing (PT) revealed some cracks on the inner surface of the component. The cracks are mostly found to be initiated from the heat affected zone or HAZ of the weldments. The cracks were extended in perpendicular to the weld in the head side (**Figure 9**). Metallographic investigation showed that all the cracks had propagated essentially in transgranular mode, and they had the appearance of branching as it is shown in **Figure 10**. These features indicate occurrence of the stress corrosion cracking on the reactor [44].

7.2.1 Material, environment, and source of stress

The reactor shell is made of AISI 316L stainless steel and operates at a pressure of 4.5 MPa. The operating temperature is 200°C. The processing environment consists of some kind of neutral organic compound and hydrogen. The base metal and corrosion products were analyzed by scanning electron microscope (SEM) and energy dispersive X-ray (EDAX) analyzer. The results indicated that the reactor was made of AISI316L stainless steel. The analysis also revealed the presence of chlorides in corrosion products. The equipment history showed that the reactor was not annealed after weld though the thickness of the reactor was 45 mm. Therefore, remaining residual stress in services arise to a level of ultimate strength.

7.2.2 Mechanism

The reactor has a thickness of 45 mm. Welding was used in manufacturing processes. A high magnitude of the welding residual stress is created (in

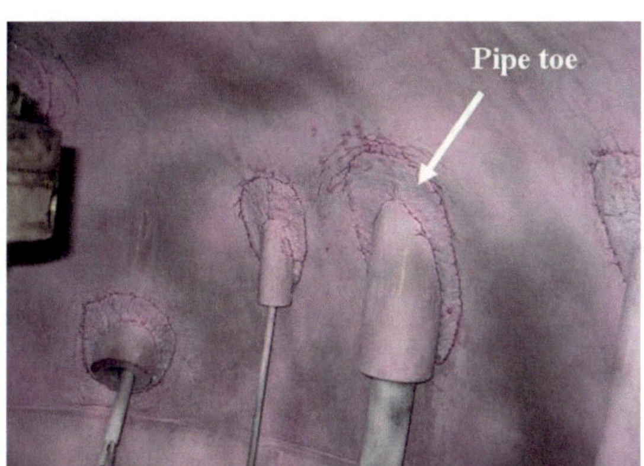

Figure 9.
Cracks in heat affected zone of the weld toe.

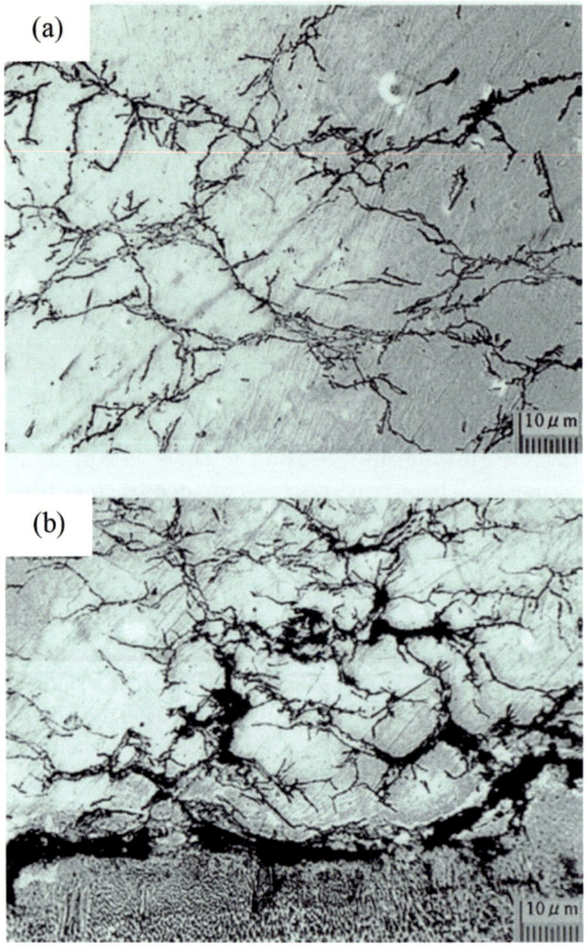

Figure 10.
The stress corrosion cracking: (a) heat affected zone and (b) the weld line.

magnitude of the ultimate strength) during welding operation. The chloride ions came from the catalysts. For the utilized catalyst, the chloride concentration is over 60 ppm. The operating temperature of the equipment is 200°C. Investigations have shown a chloride concentration of 10 ppm is enough to motive SCC in an austenitic stainless steel grade if the metal temperature is over 60°C [2]. Therefore, chloride SCC is responsible for the failure of the reactor.

7.2.3 Prevention

- The main cause of failures is the presence of high concentration of chlorides. Therefore, decreasing the concentration of this aggressive agent in the catalyst can be an effective approach. Experience on the other reactor used in the same operational conditions shows that the concentration of chloride in the catalyst should be less than 60 ppm for the AISI 316L material.

- Selecting more resistant materials to SCC. Since AISI 316L cannot tolerate such a high concentration of chloride, more corrosion-resistant alloys, such as duplex stainless steel, should be utilized if the concentration of the chloride in the catalyst cannot decrease to a proper level.

- Decreasing the magnitude of stress. As welding residual stress is a main source of stress attributed to the stress corrosion cracking, decreasing its order should decline the possibility of SCC. It means that the reactor should be stress relived after welding.

7.3 Case 3: stress corrosion cracking of a copper refrigerant tubing

Copper alloy C12200 tubes of a refrigerant failed in bend area (**Figure 11**). Failures were occurred only after approximately 6 months of services. The cracked area is shown in **Figure 12**. SEM image of fractured areas revealed the branch cracked through the microstructure, **Figure 13**, which is characteristic of a stress corrosion cracking phenomenon [61].

7.3.1 Material, environment, and source of stress

Chemical analysis of the tube materials proved the C12200 copper alloy composition. Analysis of the black foam insulation utilized with the subject copper tubing revealed the presence of ammonia by reason of the fabrication process. The stress comes from the cold-working of copper tubes during manufacturing processes.

7.3.2 Mechanism

The phosphorized copper alloy tube failed due to SCC induced by a moist ammonia environment related the black foam insulation. Although the resistance of copper alloy under dry ammonia environment is notable, gathering of moisture ease by the bent tube geometry provokes the tubing to be exposed to wet ammonia, which is substantially more. The sensitivity of the copper tubing to SCC was

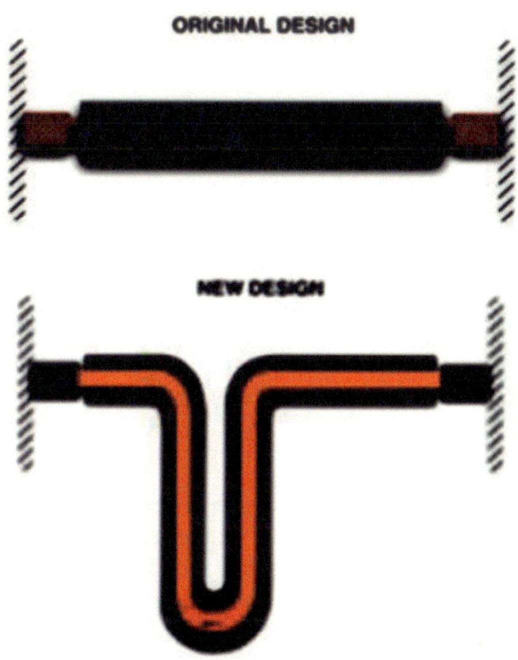

Figure 11.
New and old design of refrigerating system.

Figure 12.
Macroscopic image cracked copper tube.

Figure 13.
SEM image of the intergranular crack growth in the copper tube.

enhanced in consequence of the cold work operation for bending the tube at an angle of 180°. Stress cycling and/or repeated loading increased the sensitivity to ammonia stress corrosion cracking as a result of the service-induced oscillational loading from the other components linked to the tube.

7.3.3 Prevention

Cold work operation generated high magnitude of residual stress which is responsible for the tube failure. Hence, it is suggested that the copper tubing be annealed after the bending operation in order to eliminate the adverse impacts of the cold working and also decrease the residual stresses available for ammonia stress corrosion cracking. Another recommendation is to avoid the installation of the tube under stress condition.

8. Conclusion

Stress corrosion cracking failure is because of combined interaction of sensitizes material, tensile mechanical stress, and corrosive environment. Chloride stress corrosion cracking of austenitic steels and seasonal cracking of copper alloys are classical example of SCC. During stress corrosion cracking, the metal or alloy is basically uncorroded over most of its surface, while fine and branch cracks growth into the bulk of material and lead the components to unpredictable premature failures. The main source of stress attributed to the stress corrosion cracking is come from the residual stress. The well-known sources of the residual stress are welding and fabrication processes such as machining and bending. The stress must be in tensile form, and comprehensive stresses can be utilized to restrict SCC. Surface defects that either preexist or are created during service due to corrosion, wear, or other processes accelerate the stress corrosion cracking phenomenon. Therefore, the role of imperfections must be taken into account. Fracture mechanics introduced a parameter named fracture toughness or $K_{Ic} = \sigma\sqrt{\pi a}$ that considered the performance of flaws in fracture behavior of materials. According to this relation, there is a critical length of crack, α, in which the materials indicate the brittle failure. The fracture toughness in the presence of corrosive environment declines considerably and displayed by $K_{Iscc.}$ It means in conditions prone to SCC, a smaller flaw size led the materials to catastrophic brittle fractures.

If one of three SCC elements does not exist, this kind of failure will not develop. Accordingly, the solving methods can be one or more of these factors:

- Lowering the stress levels
- Eliminating aggressive species in the environment
- Changing the material is one possible
- Applying cathodic protection
- Coating that is sometimes effective
- Shot-peening that produces residual compressive stress

Author details

Alireza Khalifeh
Inspection Department, Shiraz Petrochemical Complex, Shiraz, Iran

*Address all correspondence to: areza1006@gmail.com

References

[1] Ahmad Z. Principles of Corrosion Engineering and Corrosion Control. UK: Elsevier; 2006

[2] Bennett D. Continuous digester cracking task group reports on past year's efforts. TAPPI Journal of the Technical Association of the Pulp and Paper Industry. 1981;**64**(9):75-77

[3] Khalifeh A, Banaraki AD, Daneshmanesh H, Paydar M. Stress corrosion cracking of a circulation water heater tubesheet. Engineering Failure Analysis. 2017;**78**:55-66

[4] Parnian N. Failure analysis of austenitic stainless steel tubes in a gas fired steam heater. Materials and Design (1980-2015). 2012;**36**:788-795

[5] Prawoto Y, Moin A, Tadjuddin M, Nik WW. Effect of microstructures on SCC of steel: Field failure analysis case study and laboratory test result. Engineering Failure Analysis. 2011;**18**(7):1858-1866

[6] Ling J et al. Failure analysis of 304H stainless steel convection tube serviced in an ethylene cracking furnace. Engineering Failure Analysis. 2019;**97**:399-407

[7] Farrahi G, Chamani M, Kashyzadeh KR, Mostafazade A, Mahmoudi A, Afshin H. Failure analysis of bolt connections in fired heater of a petrochemical unit. Engineering Failure Analysis. 2018;**92**:327-342

[8] Xu S, Meng W, Wang C, Sun Z, Zhang Y. Failure analysis of TP304H tubes in the superheated steam section of a reformer furnace. Engineering Failure Analysis. 2017;**79**:762-772

[9] Yoon KB, Yu JM, Nguyen TS. Stress relaxation cracking in 304H stainless steel weld of a chemical reactor serviced at 560° C. Engineering Failure Analysis. 2015;**56**:288-299

[10] Huang Y, Tang X, Yu H, Qian Y, Wu R. Failure analysis of expansion bellows serving in steam pipe network. In: ASME 2013 Pressure Vessels and Piping Conference; American Society of Mechanical Engineers Digital Collection; 2014

[11] Fontana MG. Corrosion Engineering. India: Tata McGraw-Hill Education; 2005

[12] Sedriks AJ. Corrosion of Stainless Steel. USA: John Wiley & Sons; 1996

[13] Khalifeh A. Stress corrosion cracking damages. In: Failure Analysis. UK: IntechOpen; 2019

[14] Pugh E, Craig J, Sedriks A. Stress-Corrosion Cracking of Copper, Silver and Gold Alloys. 1969

[15] Kannan MB, Shukla P. Stress corrosion cracking (SCC) of copper and copper-based alloys. In: Stress Corrosion Cracking. UK: Elsevier; 2011. pp. 409-426

[16] Chen Y, Tzou R, Chang Y, Wang L, Oung J, Shih H. Two distinct fracture modes of copper alloys in fluoride environments. Corrosion Science. 2005;**47**(1):79-93

[17] El Domiaty A, Alhajji J. The susceptibility of 90Cu-10Ni alloy to stress corrosion cracking in seawater polluted by sulfide ions. Journal of Materials Engineering and Performance. 1997;**6**(4):534-544

[18] Wilmott MJ, Sutherby RL. The role of pressure and pressure fluctuations in the growth of stress corrosion cracks in line pipe steels. In: 1998 2nd International Pipeline Conference;

American Society of Mechanical Engineers; 1998. pp. 409-422

[19] Harle BA, Beavers JA, Jaske CE. Effect of stress corrosion cracking on integrity and remaining life of natural gas pipelines. In: Corrosion 96; NACE International; 1996

[20] Sridhar N, Price K, Buckingham J, Dante J. Stress corrosion cracking of carbon steel in ethanol. Corrosion. 2006;**62**(8):687-702

[21] Wang J, Atrens A. SCC initiation for X65 pipeline steel in the "high" pH carbonate/bicarbonate solution. Corrosion Science. 2003;**45**(10):2199-2217

[22] Contreras A, Hernández S, Orozco-Cruz R, Galvan-Martínez R. Mechanical and environmental effects on stress corrosion cracking of low carbon pipeline steel in a soil solution. Materials and Design. 2012;**35**:281-289

[23] Qiao L, Luo J, Mao X. Hydrogen evolution and enrichment around stress corrosion crack tips of pipeline steels in dilute bicarbonate solution. Corrosion. 1998;**54**(2):115-120

[24] Zhang X-Y, Lambert S, Sutherby R, Plumtree A. Transgranular stress corrosion cracking of X-60 pipeline steel in simulated ground water. Corrosion. 1999;**55**(3):297-305

[25] Parkins RN. A review of stress corrosion cracking of high pressure gas pipelines. In: Corrosion 2000; NACE International; 2000

[26] Popović M, Romhanji E. Stress corrosion cracking susceptibility of Al–Mg alloy sheet with high Mg content. Journal of Materials Processing Technology. 2002;**125**:275-280

[27] Speidel MO. Stress corrosion cracking of aluminum alloys. Metallurgical Transactions A. 1975;**6**(4):631

[28] Rao AU, Vasu V, Govindaraju M, Srinadh KS. Stress corrosion cracking behaviour of 7xxx aluminum alloys: A literature review. Transactions of Nonferrous Metals Society of China. 2016;**26**(6):1447-1471

[29] Burleigh T. The postulated mechanisms for stress corrosion cracking of aluminum alloys: A review of the literature 1980-1989. Corrosion. 1991;**47**(2):89-98

[30] Tsai T, Chuang T. Role of grain size on the stress corrosion cracking of 7475 aluminum alloys. Materials Science and Engineering A. 1997;**225**(1-2):135-144

[31] Cao S, Lim CVS, Hinton B, Wu X. Effects of microtexture and Ti_3Al ($\alpha 2$) precipitates on stress-corrosion cracking properties of a Ti-8Al-1Mo-1V alloy. Corrosion Science. 2017;**116**:22-33

[32] Pilchak A, Young A, Williams J. Stress corrosion cracking facet crystallography of Ti–8Al–1Mo–1V. Corrosion Science. 2010;**52**(10):3287-3296

[33] Banerjee D, Williams J. Microstructure and slip character in titanium alloys. Defence Science Journal. 1986;**36**(2):191-206

[34] Kumosa L, Armentrout D, Kumosa M. The effect of sandblasting on the initiation of stress corrosion cracking in unidirectional E-glass/polymer composites used in high voltage composite (non-ceramic) insulators. Composites Science and Technology. 2002;**62**(15):1999-2015

[35] Lake G. Ozone cracking and protection of rubber. Rubber Chemistry and Technology. 1970;**43**(5):1230-1254

[36] Tse M. Ozone cracking and flex cracking of crosslinked polymer blend compounds. Journal of Applied Polymer Science. 2007;**103**(4):2183-2196

[37] Cataldo F. The action of ozone on polymers having unconjugated and cross-or linearly conjugated unsaturation: Chemistry and technological aspects. Polymer Degradation and Stability. 2001;**73**(3):511-520

[38] Ma B, Andersson J, Gubanski SM. Evaluating resistance of polymeric materials for outdoor applications to corona and ozone. IEEE Transactions on Dielectrics and Electrical Insulation. 2010;**17**(2):555-565

[39] Bocanegra-Bernal M, De La Torre SD. Phase transitions in zirconium dioxide and related materials for high performance engineering ceramics. Journal of Materials Science. 2002;**37**(23):4947-4971

[40] Wiederhorn S, Bolz L. Stress corrosion and static fatigue of glass. Journal of the American Ceramic Society. 1970;**53**(10):543-548

[41] Cheng YF. Stress Corrosion Cracking of Pipelines. USA: John Wiley & Sons; 2013

[42] Kan W, Pan H. Failure analysis of a stainless steel hydrotreating reactor. Engineering Failure Analysis. 2011;**18**(1):110-116

[43] Gokhale A, Handbook ASM. Failure analysis and prevention. Vol. 11. In: Shipley RJ, Becker WT, editors. Materials Park, Ohio: ASM International; 2002. pp. 538-556

[44] Wang Y, Lu Y-B, Pan H-L. Failure analysis of a hydro-processing reactor. Engineering Failure Analysis. 2009;**16**(1):11-18

[45] Raman RS. Role of caustic concentration and electrochemical potentials in caustic cracking of steels. Materials Science and Engineering A. 2006;**441**(1-2):342-348

[46] Singh PM, Ige O, Mahmood J. Stress corrosion cracking of type 304L stainless steel in sodium sulfide-containing caustic solutions. Corrosion. 2003;**59**(10):843-850

[47] Horowitz HH. Chemical studies of polythionic acid stress-corrosion cracking. Corrosion Science. 1983;**23**(4):353-362

[48] Craig B, Lane R. Environmentally-assisted cracking: Comparing the influence of hydrogen, stress, and corrosion on cracking mechanisms. AMPTIAC Quarterly. 2005;**9**(1):17-24

[49] Revie R, Ramsingh R. Effects of potential on strees–corporsion cracking of grade 483 (X-70) HSLA line-pipe steels. Canadian Metallurgical Quarterly. 1983;**22**(2):235-240

[50] Revie RW. Corrosion and Corrosion Control: An Introduction to Corrosion Science and Engineering. USA: John Wiley & Sons; 2008

[51] Kiefer GC, Harple WW. Stress-corrosion cracking of commercially pure titanium. Metal Progress. 1953;**63**(2):74-76

[52] Beck AF, Sperry PR. Fundamental aspects of stress corrosion cracking. In: Staehle RW, Forty AJ, editors. Houston, TX: National Association of Corrosion Engineers. 1969. pp. 513-529

[53] Vishnu PR. Solid-State transformations in weldments. ASM handbook. Vol. 6. USA: ASM International; 1994

[54] Deng D. FEM prediction of welding residual stress and distortion in carbon steel considering phase transformation effects. Materials and Design. 2009;**30**(2):359-366

[55] Leggatt R. Residual stresses in welded structures. International Journal of Pressure Vessels and Piping. 2008;**85**(3):144-151

[56] Hatty V, Kahn H, Heuer AH. Fracture toughness, fracture strength, and stress corrosion cracking of silicon dioxide thin films. Journal of Microelectromechanical Systems. 2008;**17**(4):943-947

[57] Dieter GE, Bacon DJ. Mechanical Metallurgy. New York: McGraw-Hill; 1986

[58] Rihan R, Raman RS, Ibrahim RN. Circumferential notched tensile (CNT) testing of cast iron for determination of threshold (KISCC) for caustic crack propagation. Materials Science and Engineering A. 2005;**407**(1-2):207-212

[59] Ibrahim RN, Rihan R, Raman RS. Validity of a new fracture mechanics technique for the determination of the threshold stress intensity factor for stress corrosion cracking (KIscc) and crack growth rate of engineering materials. Engineering Fracture Mechanics. 2008;**75**(6):1623-1634

[60] Abedi SS, Abdolmaleki A, Adibi N. Failure analysis of SCC and SRB induced cracking of a transmission oil products pipeline. Engineering Failure Analysis. 2007;**14**(1):250-261

[61] McDougal J, Stevenson M. Stress-corrosion cracking in copper refrigerant tubing. Journal of Failure Analysis and Prevention. 2005;**5**(1):13-17

Section 2

Failure Analysis and Durability Issues

Chapter 4

The Position and Function of Macroscopic Analysis in the Failure Analysis of Railway Fasteners

Guodong Cui, Shuaijiang Yan, Chengsong Zhang, Dazhi Chen and Chuan Yang

Abstract

Macroscopic analysis plays an important role in failure analysis, which cannot be replaced by other analyzing methods. In recent years, with the development of characterization techniques, more and more engineers and technicians rely on the advanced analytical testing methods in the process of failure analysis, ignoring the methods and means of macroscopic analysis. This can easily lead to some wrong judgments. Therefore, this chapter will combine with the cases to explain the position and role of macroanalysis in the failure analysis of rail fastening clips and to offer references for engineers and technicians in relevant fields.

Keywords: failure analysis, macroscopic analysis, railway fastener, fracture, crack initiation

1. Introduction

Macroscopic analysis refers to the method of observation, description, and analysis of the macroscopic features, such as shape, morphology, dimensional accuracy, cracks, processing defects, fracture surface, etc., of materials by the naked eye or using a magnifier at a low magnification (usually less than 50 times magnifying) [1–3]. Due to its simplicity and convenience, macroscopic analysis is widely used in the production and engineering practice. However, in recent years, with the development of the material characterization techniques and equipment, more and more engineers and technicians are inclined to rely on the advanced characterization equipment in the actual analysis and testing process, thus ignoring various macroscopic analysis methods. Particularly in the failure analysis of actual working parts, if due attention failed to be paid to macroanalysis, some wrong judgments can be easily made, which will eventually lead to the catastrophic consequences [4, 5].

In the process of failure analysis, macroanalysis is usually the first and the most important step. Through the macroanalysis, the failure mode, such as wear, corrosion, severe plastic deformation or fracture, etc., can be determined rapidly. In addition, the specific location of the failing point in the entire component can be determined by macroanalysis, such as whether the failed position bear the maximum force, whether the stress is concentrated at specific locations, and whether a processing defect exists near the fracture surface, etc. These judgments are helpful to find the specific cause of the failure.

The application of macroscopic analysis in metal materials mainly consists of etching, imprinting, nondestructive testing, and fracture surface observation methods [6]. Among them, the etching and imprinting methods are mainly used for detecting metallurgical defects such as microstructural segregation, inclusion, looseness, and pores in metal parts and are also used for cleaning the fracture surface of the failed samples [7, 8]. However, with the continuous improvement of manufacturing processes and technologies in recent years, the metallurgical defects in metal parts have been greatly reduced, and the failure of metal parts is rarely caused by metallurgical defects. Therefore, the application of etching and imprinting methods in failure analysis became fewer. Comparing to the etching and imprinting methods, nondestructive testing technique is an important means for detecting the sample surface and subsurface or internal defects without spoiling the metal parts. It is often used for testing sample quality and assisting the failure analysis process [9–11].

In the actual failure cases, fracture failure is the most important failure mode of mechanical parts. Therefore, the fracture surface observation method plays an important role in the failure analysis and is one of the most important and commonly used methods in the failure analysis process [12]. The stress condition and the failure process can be judged through observing the position of the fracture surface. By observing the characteristics of the fracture surface, the position of the crack source can be accurately determined [13], which provides an important basis for further analysis of the causes of subsequent fractures.

Railway fasteners, used for connecting the rails with the roadbed and playing a role of shock absorption, are important working parts in the railway transportation [14–16]. It will seriously affect the safety of the train if the fracture occurs. Based on the railway fastener cases failed in different ways, this chapter reveals the causes of the fracture from the perspective of macroanalysis and discusses the position and role of macroanalysis in the failure analysis.

2. Macroanalysis of the railway fastener fracture cases

2.1 The premature fatigue failure of fasteners caused by processing defects

Figure 1 shows the fracture condition of a rail fastening clip after the fatigue test of 4 × 105 cycles. However, according to the TB/T2329-2002 Chinese standard for the fatigue tests of rail fasteners, the samples should not fracture after 5 × 106 cycles. In order to find out the reasons of the premature failure, macroscopic analysis was conducted on the failed sample. As shown in **Figure 1a**, the fracture surface is not located in the position bearing the maximum stress during fatigue tests. The detailed observation reveals that a processing defect exists near the crack initiation region, as is shown in **Figure 1b**. This defect is mainly caused by the excessive extrusion of the hot plastic mold during induction heating, resulting in a local stress concentration at the defect. Subsequent quenching further increased the local stress at defects. During the fatigue tests, crack initiated from the stress concentrated position at the defects leading to the premature failure of the tested sample. As shown in **Figure 1c**, the characteristics of the fracture surface were revealed by the macroscopic analysis. The fracture surface shows the feature of a typical fatigue failure mode. The crack initiation region, crack expansion zone, and the final fracture region can be clearly observed on the fracture surface as marked in **Figure 1c**. Since the crack expansion zone occupies a large area, it indicates that,

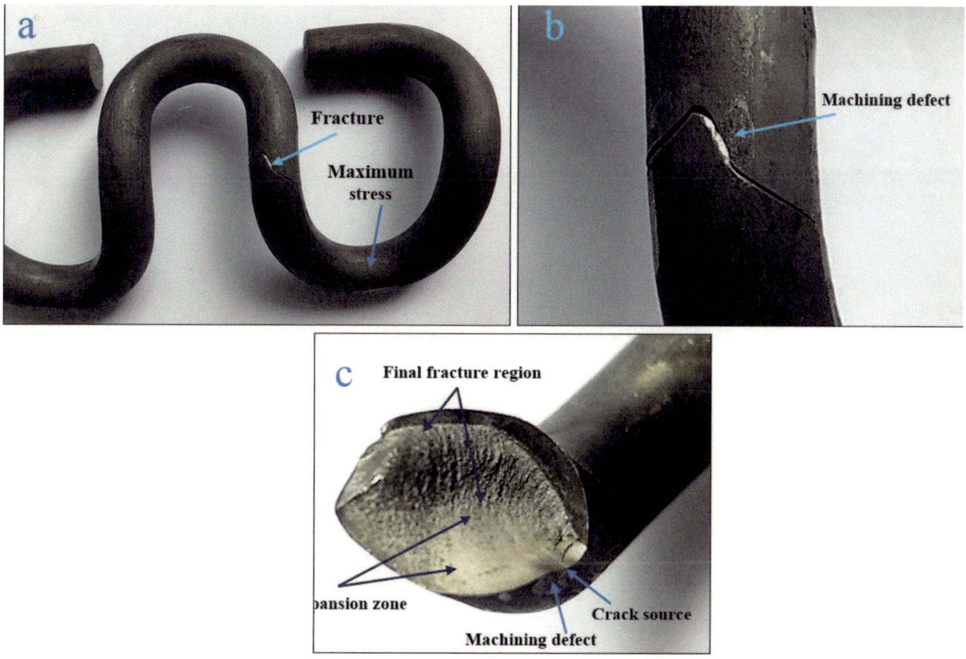

Figure 1.
Macroscopic analysis of the premature fatigue fractured rail fastening clip caused by surface processing defects: (a) the overall morphology of the fractured clip, (b) the processing defects near the crack initiation region, and (c) the morphology of the fractured surface.

after the cracks initiated at the surface defect, the sample undergoes a crack expansion period under the cyclic load before the final fracture. According to the theoretical equation reported in the literature [17, 18], the fracture strength of the tested rail fastening clip can be calculated based on the area of the instantaneous fracture zone and the maximum load applied.

In addition to the macroscopic analysis of the failed sample itself, analysis of the manufacturing environment, production conditions, and the service environment of the samples is also required. Sometimes, the external environments and service conditions can also play a key role on the sample failure. For example, oil was used as a common cooling medium in the heat treatment for many alloy steels. When the humidity in the heat treatment plant is high, the content of water in the quenching oil will increase continuously with the increasing of time. Therefore, the cooling rate will increase significantly when quenching is carried out in the oil with a certain concentration of water. When quenching alloy steels with a good hardenability, the internal stress will increase greatly due to the higher cooling rate, which leads to the increased risk of cracking after quenching of the sample [19].

2.2 The premature failure caused by improper heat treatments

Another example shows the effect of the producing process on the failure conditions of the rail fasteners. In order to improve production efficiency and reduce cost, the rail fasteners are subjected to medium frequency induction heating treatment before being deformed into the "M" shape. Then, the railway fasteners were quenched by the residual heat of induction heating after the thermoforming. In order to further reduce costs and improve efficiency, a company replaced the original intermediate frequency induction heating with high frequency induction heating in the production of railway fasteners. Because the heat generated by the high-frequency induction

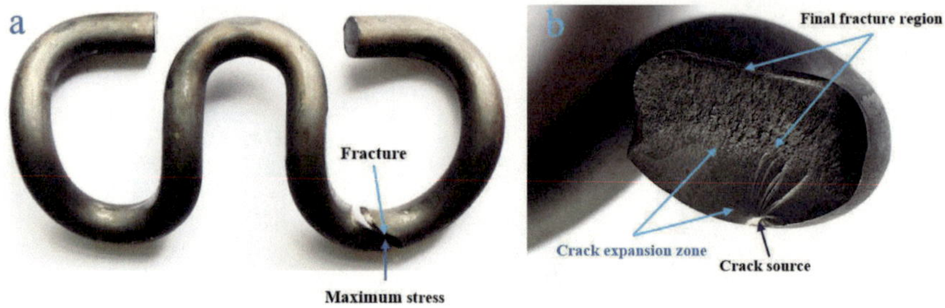

Figure 2.
The macroscopic morphology of the premature fatigue fracture samples processed by the high-frequency-inducing heating.

heating is more concentrated on the sample surface, it is easy to cause the uneven heating of the samples, resulting in large residual stress. In addition, the macroscopic factors such as insufficient heating time, insufficient heating power, or incorrect heating location of the sample may cause the inhomogeneity of the microstructure in the heat-treated samples. This type of microstructural inhomogeneity will lead to the property difference in certain regions, which can greatly reduce the fatigue life of the railway fasteners and cause the premature fatigue failure. **Figure 2** shows the fracture condition of railway fasteners, and the fatigue test is carried out after the high-frequency-inducing heating process. The premature failure occurred after 1.2 × 106 fatigue cycles (normally 5 × 106 cycles without fracture).

Compared with the case shown in **Figure 1** (case 1), the fracture position of the fastener is at the region bearing the maximum stress (as marked by the arrow in **Figure 2a**). The appearance of the failed region is normal, no obvious processing defect can be observed, and the fracture morphology also has typical fatigue fracture morphology including the crack initiation region, crack expansion zone, and the final fracture region [20, 21]. The area of the crack expansion region is comparable to that of case 1, but its fatigue life is much higher than the case 1, indicating that the fatigue crack growth rate is significantly lower than case 1. However, because the sample still did not reach the fatigue life of the typical fastener, it belongs to the abnormal fracture type. Different from case 1, the reason of the premature failure of case 2 cannot be directly found from the macroscopic analysis, and the microscopic analysis is therefore required. Since the fastener was produced by high-frequency heating rather than the original intermediate frequency heating, it is suspected that the uneven distribution of microstructure caused the premature fatigue fracture. Therefore, the metallographic microstructure analysis was carried out on the normal and the prematurely failed fastener at the position bearing the maximum stress during fatigue test. The position for extracting the sample parts and the corresponding microstructure are shown in **Figure 3**.

As shown in **Figure 3a** and **b**, the samples for microstructure observation of the normal and prematurely failed fasteners were cut from the same position near the fracture surface. The samples were ground by SiC paper and polished. The Nital solution (4% alcohol solution of nitric acid) was used as the etching solution. An optical microscope was used to complete the microstructure observation of the samples, and the results are shown in **Figure 3c–f**. **Figure 3c** and **d** shows the microstructure of the normal railway fastener after quenching and tempering in the medium temperature range; the typical tempered troostite and a small amount of ferrite can be clearly observed. The troostite and ferrite grains are fine, and they distributed evenly in the microstructure. **Figure 3e** and **f** shows the microstructure of the prematurely fractured samples. The obvious microstructure segregation

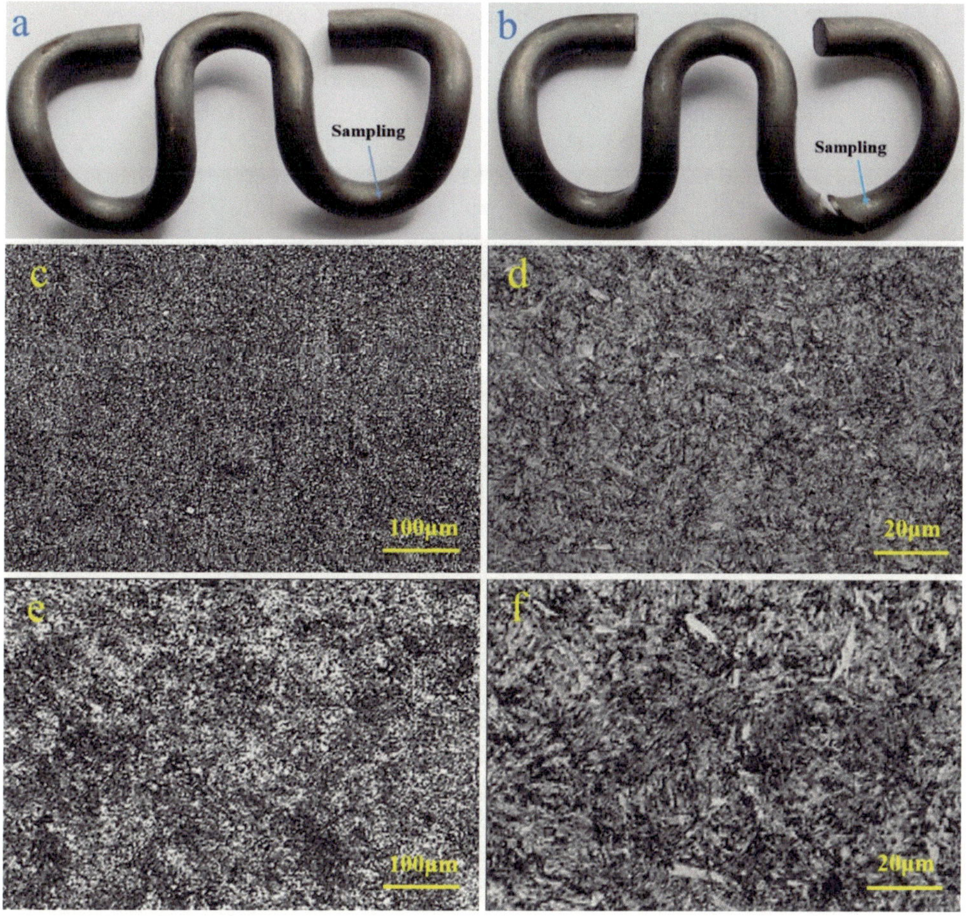

Figure 3.
The sample extraction position of normal (a) and prematurely failed railway (b) fasteners. (b, c) and (e, f) are the corresponding metallographic microstructure of the normal and prematurely failed fasteners, respectively.

can be observed from **Figure 3e**, which is caused by the large amount of dissolved cementite due to the fast heating rate of the high-frequency-inducing heating. The cementite came from the pearlite before heat treatment. As can be seen from **Figure 3f**, the grains are coarser, and in addition to the undissolved cementite, large block-shaped ferrite can also be observed. This is mainly due to the high heating rate and high temperature caused by the high-frequency-inducing heating. The rapid heating speed leads to the existence of a large amount of undissolved cementite. The heating temperature is too high, resulting in the formation of coarse microstructures [22]. This microstructural inhomogeneity can seriously affect the fatigue performance of the samples. Cracks are more prone to initiate at the microstructure with poor mechanical properties under the applied cyclic load, resulting in the premature fatigue fracture. This example shows that in the process of failure analysis, in addition to the analysis of the macroscopic characteristics of the sample, the macroscopic factors such as the production environment and the service environment of the sample parts are also important and sometimes are important causes of sample failure. It is sometimes difficult to directly and accurately determine the cause of failure from macroanalysis. This requires a combination of macroanalysis and microanalysis to achieve the accurate failure analysis results.

2.3 The effect of the service condition on the failure of railway fasteners

In addition to the manufacturing processes, the service environment is sometimes critical to the macroanalysis of failures. For example, in the following case, an e-type fastener widely used in the subway track has a premature fatigue fracture when it has been used in practice for about 1 year (the designed service life is 10 years). This kind of fastener breakage will cause major safety hazards to the safe operation of the train. It is necessary to analyze the causes of the failure in order to eliminate potential dangers in time and ensure the safe operation of the train. In order to analyze the cause of the fracture, a macroscopic observation of the broken subway track fastener clip was first carried out; the results are shown in **Figure 4**.

As can be seen from **Figure 4a**, the breaking position located at the root of the straight section of the fastener, and this straight section is installed in the fixed slot. The fracture happened at the boundary between the straight section and the residual curved part, in which the maximum stress is loaded in the broken position in the actual working condition. A detailed observation of the sample surface reveals obvious wear marks. Moreover, the wear marks are in the same direction along the transverse arc of the fastener sample (see **Figure 4b**). As can be observed from the magnified view shown in **Figure 4c**, the wear marks have a certain depth, and their propagation direction is consistent. Based on the direction and depth of the wear scar, it can be preliminarily concluded that the wear scars on the surface of the straight section of the railway fastener were caused by the relative rotation with the spring clip slot it contacted with. However, in the normal circumstances, this type of rotation is not allowed, because the rotation will reduce the pressure between the fastener and the railway track, affecting the train safety. Combined with the relatively deep wear marks, the broken fastener is subjected to a large external force before failure. The reason causing this large external force can only be determined by the on-site investigation with the understanding of the service situation.

According to the investigation of the service scene, the subway operates along a circle line, and the fasteners with premature fatigue failure occurred mostly at or near the curve region of the railway track. This indicates that in addition to the force of caused by the vibration of the track when a train passed, the spring bar is also subjected to the centrifugal force when the train is adjusting directions. Under the combined force of the vibration and the centrifugal force, the fastener rotated relatively in the slot, resulting in a surface with a consistent direction of wear scar. After multiple friction, the wear marks at the position where the fastener is in contact with the edge of the slot became deeper and deeper, and thus fatigue cracks were

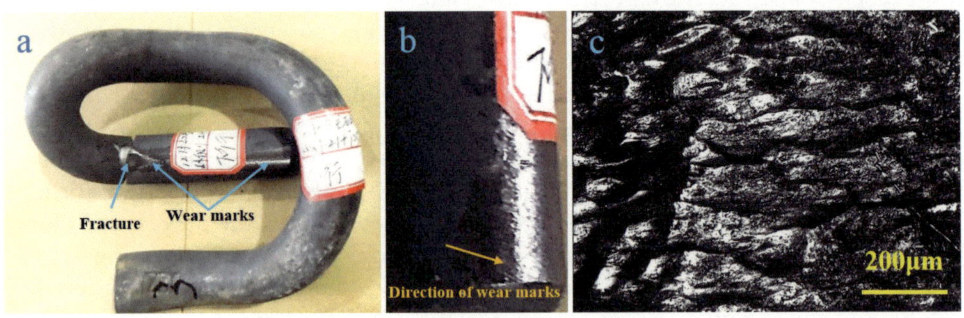

Figure 4.
The macroscopic morphology and surface wear scars of the fractured metro fastener: (a) the macroscopic morphology of the failed fastener showing the position of the fracture surface and the wear marks, (b) the obvious wear scar that can be observed on the surface of the straight section of the fastener clip, and (c) the enlarged view of wear scar.

DOI: http://dx.doi.org/10.5772/intechopen.89262

Figure 5.
The macroscopic morphology of the fracture surface and the microstructure of the samples near the fracture surface: (a) the fracture surface morphology, (b) the cross-sectional microstructure of the sample near the fracture surface, and (c) the microstructure of core region of the sample near the fracture surface.

generated. Due to the large external force, the crack spreads rapidly, resulting in the final premature fracture failure of the fastener. In order to more accurately determine the cause of the failure of the elastic strip, combined with the macroscopic analysis of the fracture surface, the metallographic microstructures near the fracture surface were further observed and analyzed. The results are presented in **Figure 5**.

From the macroscopic analysis of the fracture surface as shown in **Figure 5a**, the crack initiation region has a certain area, and the color is dark blue and slightly black, which is mainly due to the inconsistent deformation of the elastic strips on both sides of the crack, and the oxidation of the fracture surface resulted from the temperature rising caused by the relative extrusion and friction. This again proves that the elastic strips underwent severe torsional deformation during actual service. From the morphology of the crack extension area in **Figure 5a**, the extended area has obvious macroscopic fan-shaped stripes, and the spacing between the strips is large. The spacing between the fan-shaped stripes is large, which indicates that under the load of the torsion force, the severe wear scars were first produced on the surface where the stress reaches the maximum value. Then, micro-cracks were formed in the most severely worn areas. Under the combined load caused by the vibration of the rail tracks and the centrifugal force, the cracks expanded rapidly, resulting in the final premature fracture. At the same time, samples near the fracture surface were extracted for the preparation of the metallographic sample. The surface and center microstructure of the surface of the prepared samples is shown in **Figure 5b** and **c**. It can be observed that the tempered troostite (or tempered torsite) and a small amount of ferrite are the main microstructure of the elastic stripe after quenching with medium temperature tempering, which is the normal microstructure. A layer of anticorrosion treatment can be observed on the surface, which is a conventional treatment for railway fasteners.

The early failure of the railway fasteners caused by the external service environment indicates that in addition to considering the material selection and processing technology and performance of the product, the actual working environments should also be taken into consideration. In order to ensure the operation safety of the equipment, the design and manufacture of the product can be improved based on the actual working condition of the components. Only then can the safety factor in the actual service process be increased.

3. The function of macroscopic analysis in the failure analysis of railway fastening clips

Based on the above analyses of several failure cases of the railway fasteners, it can be concluded that the macroscopic analysis plays a key role in the process of failure analysis. Combined with the macroscopic characteristics of the sample, the failure site, the background data, the service environment, and other macro factors, the mode and cause of the failure can be preliminarily judged after careful observation and analysis. On the basis of macroanalysis, with the help of modern analytical test methods, microanalysis, and computer simulation techniques, the failure mechanism can be further explored to accurately determine the cause of failure. Moreover, the solutions can be given based on the causes of the failure for avoiding the occurrence of disasters.

In Section 2.1, if the macroscopic analysis is not carried out, the processing defects near the crack source cannot be found, and the causes of the premature fracture of the fasteners will not be accurately determined, which will have a serious impact on the production and sales of the products. If the unqualified products with surface defects flow into the market, it will cause serious danger to the safety of the train operation. Moreover, as discussed in Section 2.2, combined with the macroscopic factors of the changes of production conditions (medium-frequency induction heating to high-frequency induction heating), it is preliminarily believed that the premature fatigue failure of the railway fastener is caused by the uneven heating. Then, under the guidance of the macroanalysis, the microscopic analysis is carried out, and the cause of the premature fatigue failure is finally determined. Based the failure analysis results, the manufacturers were told to make improvements in time to avoid major economic losses and safety hazards. Similarly, in the example of Section 2.3, the causes of the fastener premature fracture were determined by combining the macroscopic damage on the appearance of the failed fastener with the complex force during service, through the investigation of the on-site service environment. It provides an important basis for the further improvement of the performance of railway fasteners.

All in all, macroanalysis is the first and most important step in failure analysis. Firstly, through macroanalysis, the position of the failure and the failure mode can be accurately determined, which lays a foundation for further analyzing the failure causes. Secondly, with the understanding of the failure location and the failure mode, the fracture reasons can be determined based on the force analysis, the service environment, and the manufacturing processes. Finally, combined with the verification experiments, microanalysis methods and computer simulation, the causes of the failure can be accurately determined, and solutions and suggestions can be given to eliminate potential safety hazards and avoid disasters.

With the development of modern technology and manufacturing equipment, product defects caused by design, materials, and processing technology are becoming less and less, and accidents caused by the product failure are also declining year by year. However, the assembly of components, special service environment, sudden natural disasters, etc. will seriously affect the operational safety of various types of equipment and facilities, in which special attention should be paid in the failure analysis. In particular, in recent years, with the development of science and technology, a large number of high-performance new materials and products are widely used, which brings new challenges to failure analysis. It is also necessary to continuously develop and innovate failure analysis methods, means, and concepts and lay the foundation for adapting to accurate failure analysis in the new situation and environment.

Acknowledgements

The authors gratefully acknowledge financial support from the Sichuan Education Reform Project (No. JG2018-158).

Author details

Guodong Cui*, Shuaijiang Yan, Chengsong Zhang, Dazhi Chen and Chuan Yang
School of Materials Science and Engineering, Southwest Jiaotong University,
Chengdu, China

*Address all correspondence to: gdcui@swjtu.edu.cn

References

[1] Zhang T, Bao R, Fei B. Load effects on macroscopic scale fatigue crack growth path in 2324-T39 aluminium alloy thin plates. International Journal of Fatigue. 2014;**58**:193-201. DOI: 10.1016/j.ijfatigue.2013.04.014

[2] Tang XS, Wei TT. Microscopic inhomogeneity coupled with macroscopic homogeneity: A localized zone of energy density for fatigue crack growth. International Journal of Fatigue. 2015;**70**:270-277. DOI: 10.1016/j.ijfatigue.2014.10.003

[3] García-Martínez M, García de Blas Villanueva FJ, Valles González MP, Pastor Muro A. Failure analysis of the rod-end bearing of an actuating cylinder. Engineering Failure Analysis. 2019;**104**:292-299. DOI: 10.1016/j.engfailanal.2019.06.006

[4] Vukelic G, Brcic M. Failure analysis of a motor vehicle coil spring. Procedia Structural Integrity. 2016;**2**:2944-2950. DOI: 10.1016/j.prostr.2016.06.368

[5] Guo R, Xue S, Zheng L, Deng A, Liu L. Fracture failure analysis of DY08 aluminum alloy elastic coupling. Engineering Failure Analysis. 2019;**104**:1030-1039. DOI: 10.1016/j.engfailanal.2019.06.074

[6] Vander Voort GF. Metallography and failure analysis. Materials Characterization. 1994;**33**:193. DOI: 10.1016/1044-5803(94)90044-2

[7] Xu J, Sun Y, Liu B, Zhu M, Yao X, Yan Y, et al. Experimental and macroscopic investigation of dynamic crack patterns in PVB laminated glass sheets subject to light-weight impact. Engineering Failure Analysis. 2011;**18**:1605-1612. DOI: 10.1016/j.engfailanal.2011.05.004

[8] You JH. Triple-scale failure estimation for a composite-reinforced structure based on integrated modeling approaches. Part 2: Meso- and macroscopic scale analysis. Engineering Fracture Mechanics. 2009;**76**:1437-1449. DOI: 10.1016/j.engfracmech.2008.10.017

[9] Dwivedi SK, Vishwakarma M, Soni PA. Advances and researches on non destructive testing: A review. Materials Today: Proceedings. 2018;**5**:3690-3698. DOI: 10.1016/j.matpr.2017.11.620

[10] Brown M, Wright D, M'Saoubi R, McGourlay J, Wallis M, Mantle A, et al. Destructive and non-destructive testing methods for characterization and detection of machining-induced white layer: A review paper. CIRP Journal of Manufacturing Science and Technology. 2018;**23**:39-53. DOI: 10.1016/j.cirpj.2018.10.001

[11] Starke P, Wu H. Use of non-destructive testing methods in a new one-specimen test strategy for the estimation of fatigue data. International Journal of Fatigue. 2018;**111**:177-185. DOI: 10.1016/j.ijfatigue.2018.02.011

[12] Kobayashi T, Shockey DA. The relationship between fracture surface roughness and fatigue load parameters. International Journal of Fatigue. 2001;**23**:135-142. DOI: 10.1016/S0142-1123(01)00149-9

[13] Park CY, Grandt AF. Effect of load transfer on the cracking behavior at a countersunk fastener hole. International Journal of Fatigue. 2007;**29**:146-157. DOI: 10.1016/j.ijfatigue.2006.01.014

[14] Li L, Thompson D, Xie Y, Zhu Q, Luo Y, Lei Z. Influence of rail fastener stiffness on railway vehicle interior noise. Applied Acoustics. 2019;**145**:69-81. DOI: 10.1016/j.apacoust.2018.09.006

[15] Wei X, Yang Z, Liu Y, Wei D, Jia L, Li Y. Railway track fastener defect

detection based on image processing and deep learning techniques: A comparative study. Engineering Applications of Artificial Intelligence. 2019;**80**:66-81. DOI: 10.1016/j.engappai.2019.01.008

[16] Cao Z, Brake MRW, Zhang D. The failure mechanisms of fasteners under multi-axial loading. Engineering Failure Analysis. 2019;**105**:708-726. DOI: 10.1016/j.engfailanal.2019.06.100

[17] Zhang J, Song Q, Zhang N, Lu L, Zhang M, Cui G. Very high cycle fatigue property of high-strength austempered ductile iron at conventional and ultrasonic frequency loading. International Journal of Fatigue. 2015;**70**:235-240. DOI: 10.1016/j.ijfatigue.2014.09.021

[18] Yu M, Wei C, Niu L, Li S, Yu Y. Calculation for tensile strength and fracture toughness of granite with three kinds of grain sizes using three-point-bending test. PLoS ONE. 2018;**13**:e0180880. DOI: 10.1371/journal.pone.0180880

[19] Ma Z, Wang Y, Ji S, Xiong L. Fatigue properties of Ti-6Al-4V alloy friction stir welding joint obtained under rapid cooling condition. Journal of Manufacturing Processes. 2018;**36**:238-247. DOI: 10.1016/j.jmapro.2018.10.006

[20] Burcham MN, Escobar R, Yenusah CO, Stone TW, Berry GN, Schemmel AL, et al. Characterization and failure analysis of an automotive ball joint. Journal of Failure Analysis and Prevention. 2017;**17**:262-274. DOI: 10.1007/s11668-017-0240-4

[21] Robin MC, Delagnes D, Logé R, Bouchard PO, Da Costa S, Monteagudo-Galindo M, et al. Thermo-mechanical fatigue behaviour of welded tubular parts made of ferritic stainless steel. International Journal of Fatigue. 2013;**54**:84-98. DOI: 10.1016/j.ijfatigue.2013.04.004

[22] Xu L, Chen L, Sun W. Effects of soaking and tempering temperature on microstructure and mechanical properties of $65Si_2MnWE$ spring steel. Vacuum. 2018;**154**:322-332. DOI: 10.1016/j.vacuum.2018.05.029

Chapter 5

Fracture Behavior of Solid-State Welded Joints

Dattaguru Ananthapadmanaban and K. Arun Vasantha Geethan

Abstract

The nature of the fractured surface gives information about the type of failure. This chapter focuses on the study of the fractured surfaces. Solid-state welding processes, such as friction welding, friction stir welding, and laser welding, have been used for welding dissimilar joints in recent times. Different combinations of materials and different welding conditions give rise to changes in the morphology of the fractured surfaces. Material combinations that have been chosen in this study are industrially useful combinations such as titanium-stainless steel and aluminum-copper. An attempt has been made to study the fractured interfaces, mainly using scanning electron microscope (SEM). In order to achieve this objective, case studies have been made use of.

Keywords: fracture, solid-state welding, welding parameters, scanning electron micrograph, mode of fracture

1. Introduction

Fracture surfaces have been studied to give information about the nature of failure. There are basically three types of fractured surfaces: ductile fracture, as defined by the cup and cone type of appearance; brittle fracture that has a cleavage type of appearance; and fatigue fracture characterized by beach marks.

Fracture in fusion welding has been studied in detail by many researchers for a few decades, but fracture in solid-state welds has been researched upon only in recent times.

Solid-state welding mainly comprises of friction welding, friction stir welding, and laser and electron-beam welding. This chapter is aimed at studying the fractured surfaces of practical cases of welding dissimilar joints.

2. Methods and materials

The authors start with friction welding since they have worked on this topic for some years now. Friction welding is a solid-state welding technique, which has been in use for the last 20 years. The main input parameters for welding are friction pressure, upset pressure, burn-off length, and speed of rotation. These parameters can be varied to give different qualities of weld joint. It is also seen that the nature of fracture and the place where the fracture occurs differ with varying friction welding parameters.

IntechOpen

2.1 Friction welding parameters-stainless steel

Low carbon steel has been friction welded with stainless steel, and fractography for various combinations of friction welding parameters has been reported. Similarly, aluminum has been friction welded with copper, and SEM micrograph has been studied to characterize the type of failure.

The same procedure has been followed for friction stir welding and laser welding. Analysis of friction stir welded joints has been done, both for ferrous welds and nonferrous welds.

Friction welding in the case of low carbon steel-stainless steel has been carried out with the following parameters (**Table 1**).

Table 2 gives the parameters for aluminum-copper friction welds.

Welding of aluminum with copper with nickel interlayer has been carried out with the following parameters as shown below in **Table 3.**

A detailed study by Ahmed et al. [1] showed that while friction stir welding 5052 with 7075 alloys, the optimum parameters used were 1400 rpm tool speed and 200 mm/min transverse speed.

S. No.	Friction pressure (MPa)	Upset pressure (MPa)	Burn-off length (mm)	RPM (rpm)
1	120	180	6	2000
2	120	180	6	1000
3	120	180	2	2000
4	120	180	2	1000
5	120	127.5	6	2000
6	120	127.5	6	1000
7	120	127.5	2	2000
8	120	127.5	2	1000
9	40	180	6	2000
10	40	180	6	1000
11	40	180	2	2000
12	40	180	2	1000
13	40	127.5	6	2000
14	40	127.5	6	1000
15	40	127.5	2	2000
16	40	127.5	2	1000

Table 1.
Friction welding parameters for Low Carbon steel-Stainless Steel combination.

S. No.	Friction pressure (MPa)	Upset pressure (MPa)	Burn-off length (mm)	Speed of rotation (rpm)
1	64	160	2	750
2	80	120	2	1000
3	64	64	2	1000
4	32	96	3	1000

Table 2.
Friction welding parameters for Aluminium-Copper combination.

Friction pressure (MPa)	Upset pressure (MPa)	Upset time (s)	Burn-off length	Spindle speed (rpm)
40	110	3	2	1500
40	130	2.5	2	1500
60	110	2	3	1500
80	95	3.5	3	1600
80	120	3	2	1600

Table 3.
Aluminum to copper with interlayer welding parameters.

3. Results and discussion

3.1 Fractography studies of low carbon steel-stainless steel friction welds

Fractography studies were done on friction welded samples; typical SEM fractographs are shown here. Typical SEM micrograph of a failure in the low carbon steel side is shown [2]. Here, we find the failure to be of a mixed mode (**Figure 1**).

The following SEM images show failure at the weld. The friction welding parameters used are given below.

The following SEM fractographies represent failure in the stainless steel side [2].

Figures 2 and **3** show fairly ductile modes of fracture.

Figures 4 and **5** show brittle mode of fracture. Hence, it can be inferred that the mode of fracture is influenced by the friction welding parameters. More detailed studies may be necessary to come to conclusions on what parameters give ductile failure and which parameters give brittle failure. Mixed mode type of failure is also obtained for certain welding parameters.

A total of 16 experiments have been carried out while friction welding low carbon steel to stainless steel; details of the parameters used are given in the previous section.

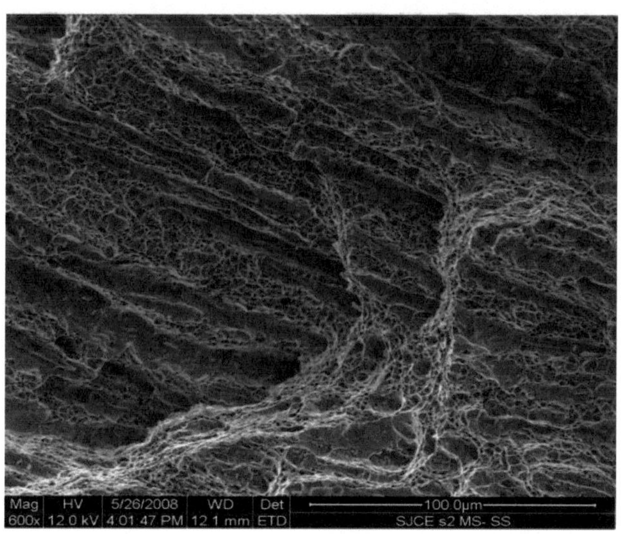

Figure 1.
Friction pressure—120 MPa, upset pressure—180 MPa, burn-off length—6 mm, and speed—2000 rpm.

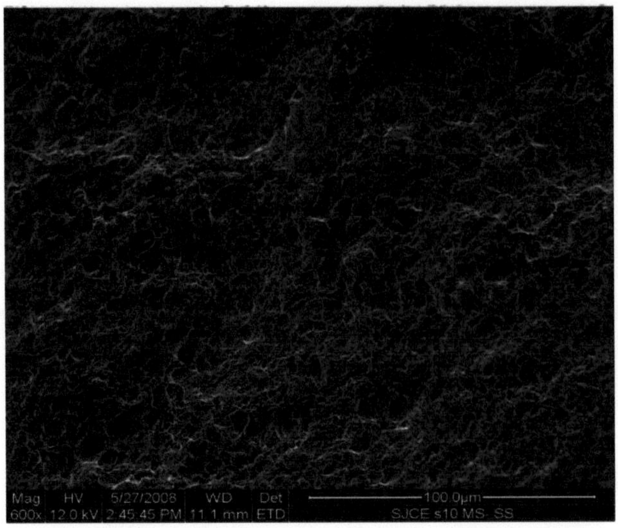

Figure 2.
Friction pressure—40 MPa, upset pressure—180 MPa, burn off length—6 mm, and speed—1000 rpm.

Figure 3.
Friction pressure—40 MPa, upset pressure—180 MPa, burn off length—2 mm, and speed—2000 rpm.

Results of the tensile testing on friction welds are presented in **Table 4**.

Specimens 2–8 failed in the low carbon steel side. Specimens 9–12 failed in the weld. Specimens 13–16 failed on the stainless steel side. Specimens 1–6 also showed a fair deal of ductile behavior. Thus, we can infer that the location of failure depends upon the friction welding parameters used. It will also be of interest to try these sort of experiments for other dissimilar metal combinations.

3.2 Fractography of aluminum-copper friction welds

Aluminum sticks to copper under certain conditions and breaks without any joining under certain different conditions. In all cases, the failure mode observed in the case of aluminum-copper was brittle fracture.

Figure 4.
F.P—40 MPa, U.P—127.5 MPa, BOL—6 mm, and speed–1000 rpm.

Figure 5.
F.P—40 MPa, U.P—127.5 MPa, BOL—2 mm, and speed—2000 rpm.

SEM studies for aluminum-copper combination were done, and some of the SEM micrographs are shown in **Figures 6–9**.

3.3 Fractography of aluminum-copper with nickel interface

In the photomicrograph of SEM-EDAX for aluminum-copper with nickel interlayer, we can observe that some copper has diffused into the interlayer (**Figure 10**). **Figure 11** shows SEM and EDAX for copper side of the weld. Here, the presence of aluminum is very marginal. It can be concluded that the nickel interlayer was effective in reducing the diffusion of aluminum to the copper side. The two zones showed here are the copper matrix and the nickel interface. The diffusion zone is between the copper and nickel with thin diffusion zone. The nickel matrix is not affected by corrosion process comparing to the copper where some pitting observed. The interface diffusion zone layer is unaffected.

Corrosion studies were also done by Ravikumar et al. on friction welded aluminum to copper with nickel interface, and SEM photographs are presented [3].

S.No.	Tensile strength (MPa)	Fracture location
1	596	Weld
2	609	Low carbon steel side
3	604	Low carbon steel side
4	625	Low carbon steel side
5	585	Low carbon steel side
6	607	Low carbon steel side
7	610	Low carbon steel side
8	632	Low carbon steel side
9	557	Weld
10	578	Weld
11	563	Weld
12	581	Weld
13	542	Weld
14	574	SS side
15	570	Weld
16	464	SS side

Table 4.
Fracture location variation for low carbon steel-AISI304L stainless steel combination.

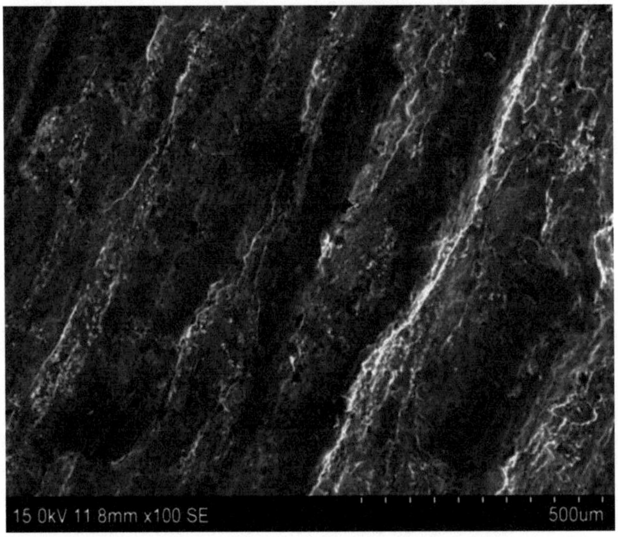

Figure 6.
F.P—64 MPa, U.P—160 MPa, BOL—2 mm, and speed—750 rpm.

Figure 12 shows SEM of the welds with nickel as interlayer after electro-chemical corrosion evaluation tests as per ASTM G 59-97.

3.4 Friction stir welding

Friction stir welding is slightly different from friction welding. It employs a rotating tool to weld two flat weld pieces. The tool chosen depends upon the material combination that is to be welded. Figure shows SEM micrograph of friction stir welded aluminum to steel joints. Ideally, solid-state welding processes, such as friction welding

Figure 7.
F.P—80 MPa, U.P—120 MPa, BOL—2 mm, and speed—1000 rpm.

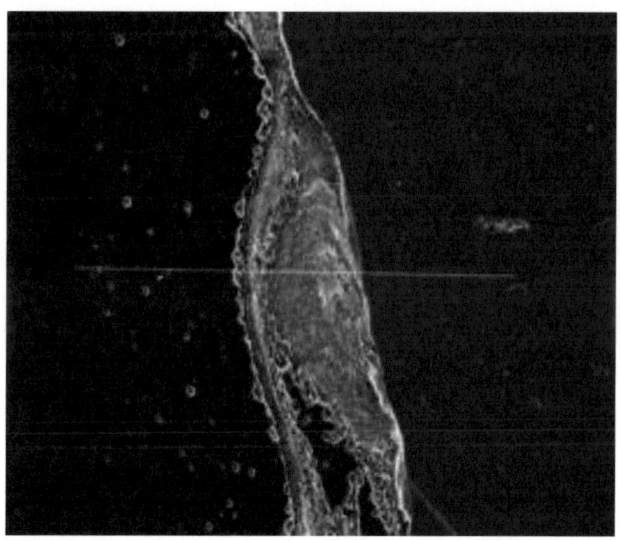

Figure 8.
F.P—64 MPa, U.P—64 MPa, and BOL—2 mm.

and friction stir welding, should not give rise to intermetallics in the weld joints, but in practice, a small amount of intermetallic is seen in the joint as shown below. These intermetallics are detrimental to the properties of the weld and efforts should be made to minimize, if not completely do way with them. This is done by varying the friction stir welding parameters and optimizing them. **Figure 13** shows the SEM photographs of aluminum steel friction stir welded joints. Intermetallic layer can be clearly seen.

3.4.1 Friction stir welding of 5 series and 7 series

These two alloys are commonly used in aircraft applications [5]. In aircraft alloys, fatigue failure is a very common mode of failure. The SEM photographs shown below clearly show striations which are an indication of fatigue failure (**Figure 14**).

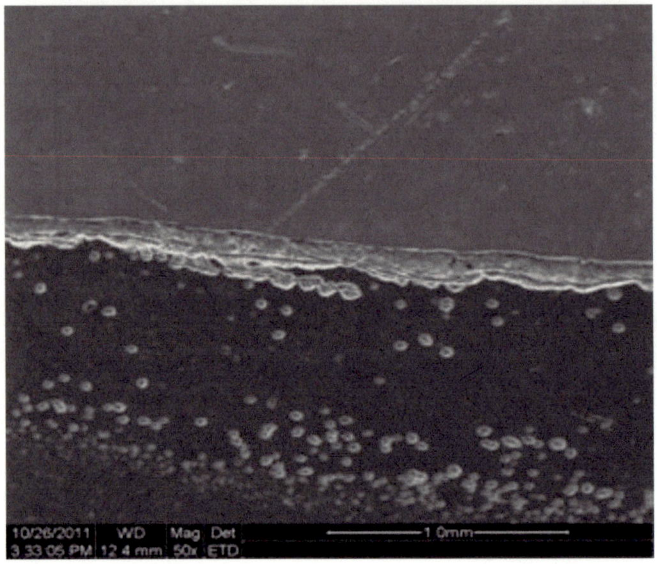

Figure 9.
F.P—32 MPa, U.P—96 MPa, and BOL—3 mm.

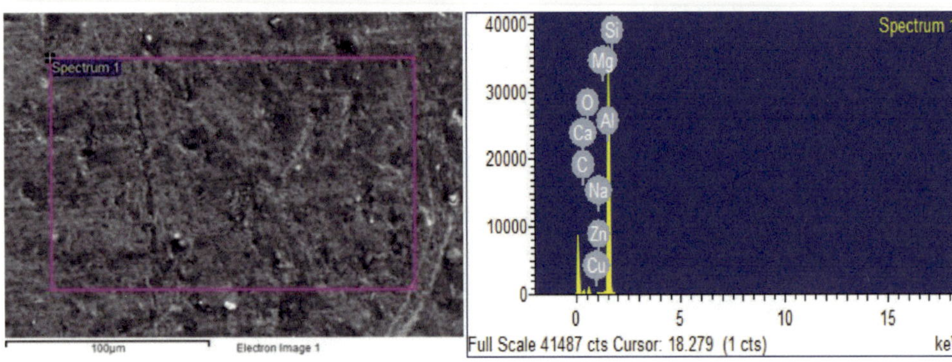

Figure 10.
Fractured surface of Al-Cu with nickel interface.

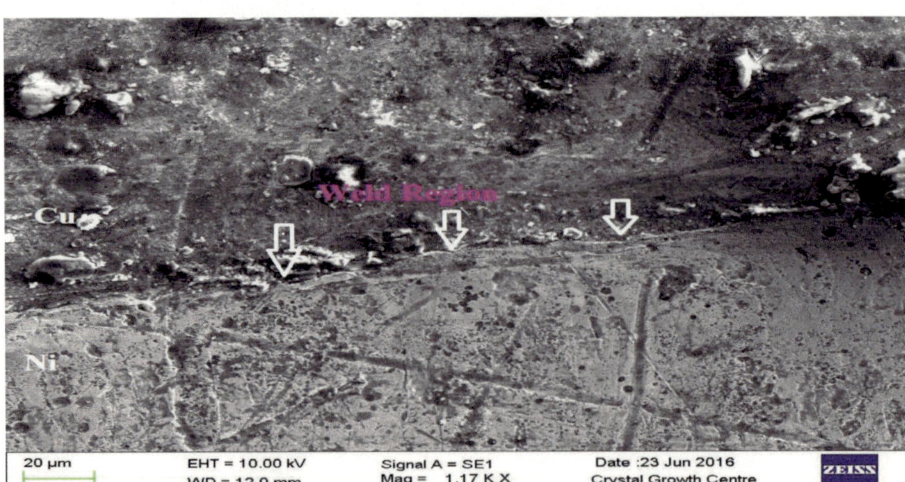

Figure 11.
SEM-EDAX on the copper side of the weld.

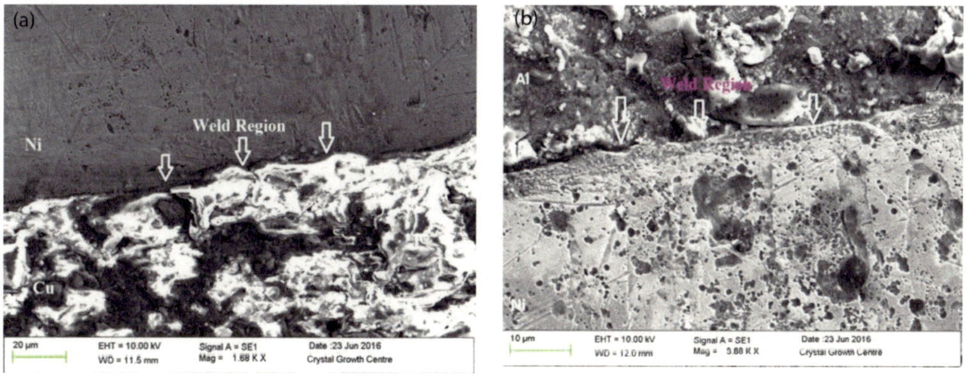

Figure 12.
Interface of specimen welded observed under FESEM at 1000 rpm: (a) copper side and (b) aluminum side.

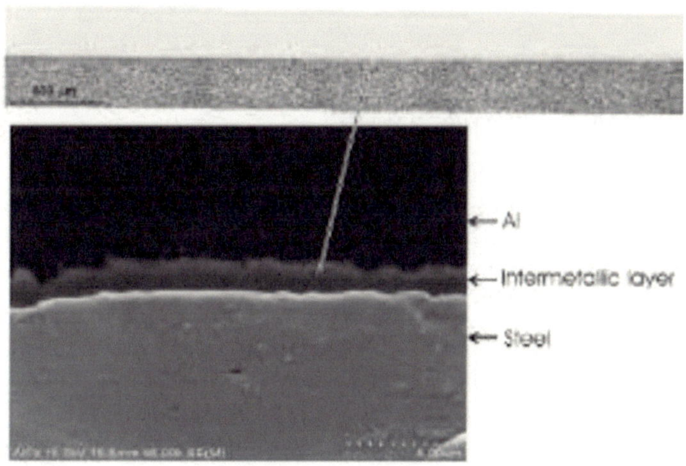

Figure 13.
SEM fractography of friction stir joints, advancing speed 200 mm/min [4].

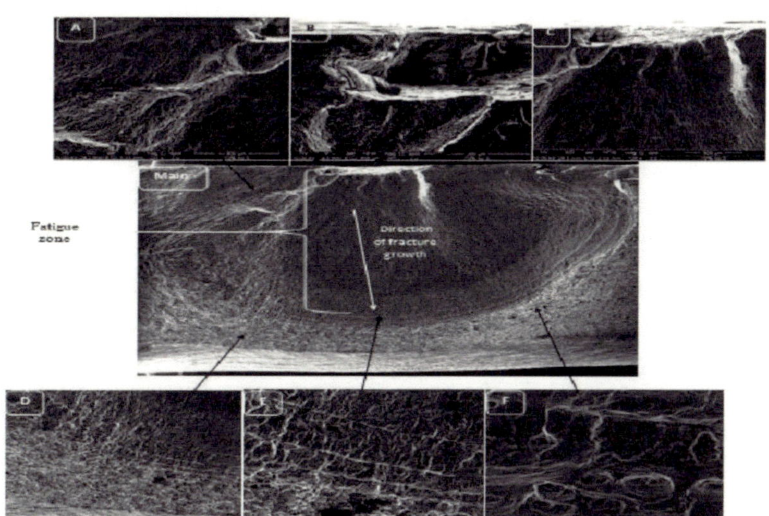

Figure 14.
SEM of 5052 and 7075 alloys—failure by fatigue [1].

It is found that the joint fabricated, using the FSW parameters of 1400 rpm (tool rotational speed) and 20 mm/min (traverse speed), showed higher strength properties compared with other joints, and the SEM shown above represents these set of parameters.

Friction stir welds, such as friction welds, can fail in parent metal or in any of the dissimilar metals. **Figure 15** shown gives the different locations of failure.

Zhang et al. studied the fractured surface of Al 6005 joints. The crack was found in the HAZ zone close to TMAZ. As shown in **Figure 16**, the fracture surface consisted of three regions. The crack initiation zone, which is located at the specimen surface, exhibited a relatively smooth microstructure (**Figure 17**). The crack propagation zone had prominent striation marks with a river-like appearance [7].

The phase of small crack initiation and propagation accounts for about 50–80% of the fatigue life. The growth rate of the major crack is similar to that of other small cracks

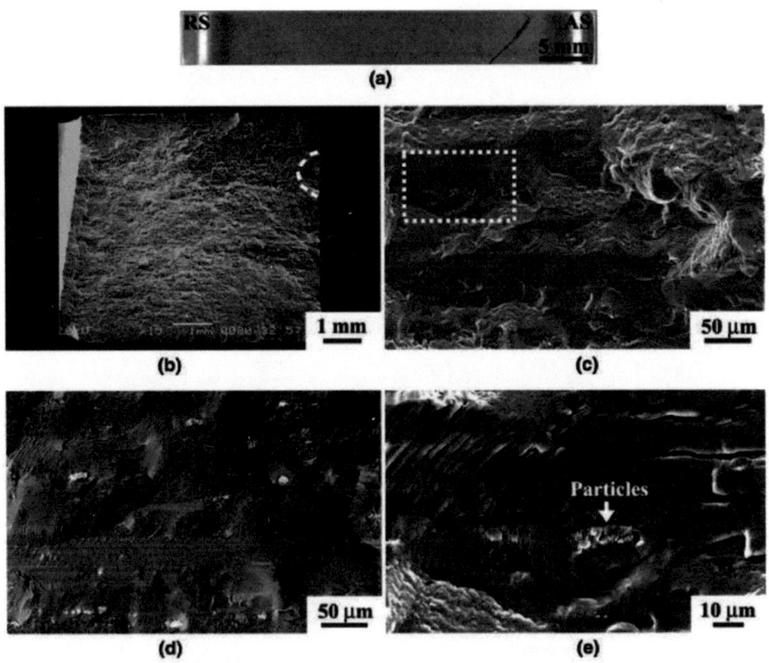

Figure 15.
Different failure locations in friction stir welds [6].

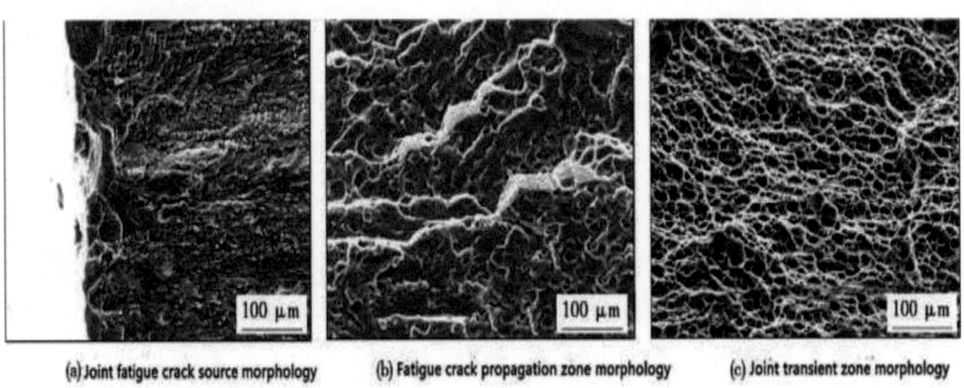

Figure 16.
Fatigue crack propagation in 6005 friction welded joints [7].

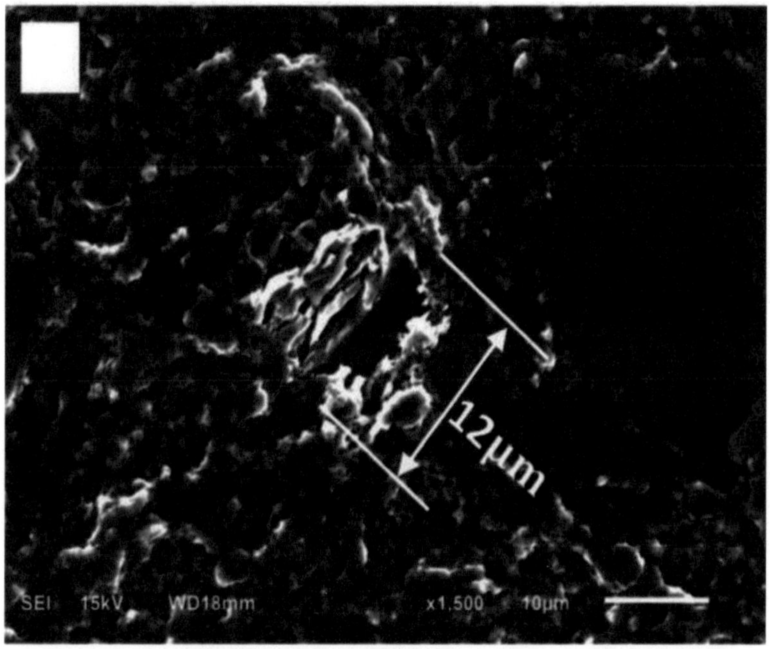

Figure 17.
Details of crack initiation during fatigue failure [8].

in the stage of small cracks growth. In the stage of long cracks growth, the growth rate of the major crack is far greater than that of other small cracks. The microstructures of different regions of the joint have an effect on the fracture mode (**Figure 18**).

3.5 Laser welded joints

Laser welding is also widely being used in recent times, the reason being laser welds have very narrow heat affected zones. This helps in defect-free welds.

3.5.1 Laser welded steels

Unlike friction welds or friction stir welds, SEM micrograph in this case shows a very smooth surface. This can be attributed to the small size of the heat affected zone.

Laser welding has been performed on high grade 960 steel joints [10].

As shown in **Figure 19**, the fracture surface near the specimen surface of BM shows a combination of equiaxed and elongated dimples, indicating that the shearing motion occurs. However, welded joints show quasi-cleavage fracture. Fracture is layered features (as shown in **Figure 19**). The tensile fracture surfaces of BM and welded joint show mostly equiaxed dimples at the center indicating ductile fracture characteristics. Cup-like dimple rupture is the main feature of the fracture surface, representing ductile type of fracture mode (**Figure 20**).

3.5.2 Laser welded nonferrous alloy

The case study taken here is the welding of aluminum alloy 6022 with zinc alloy AZ 31. Fusion zone can be clearly seen. The two base metals can also be seen. An nickel interlayer has been added in this case, which as we have seen earlier results in better corrosion resistance.

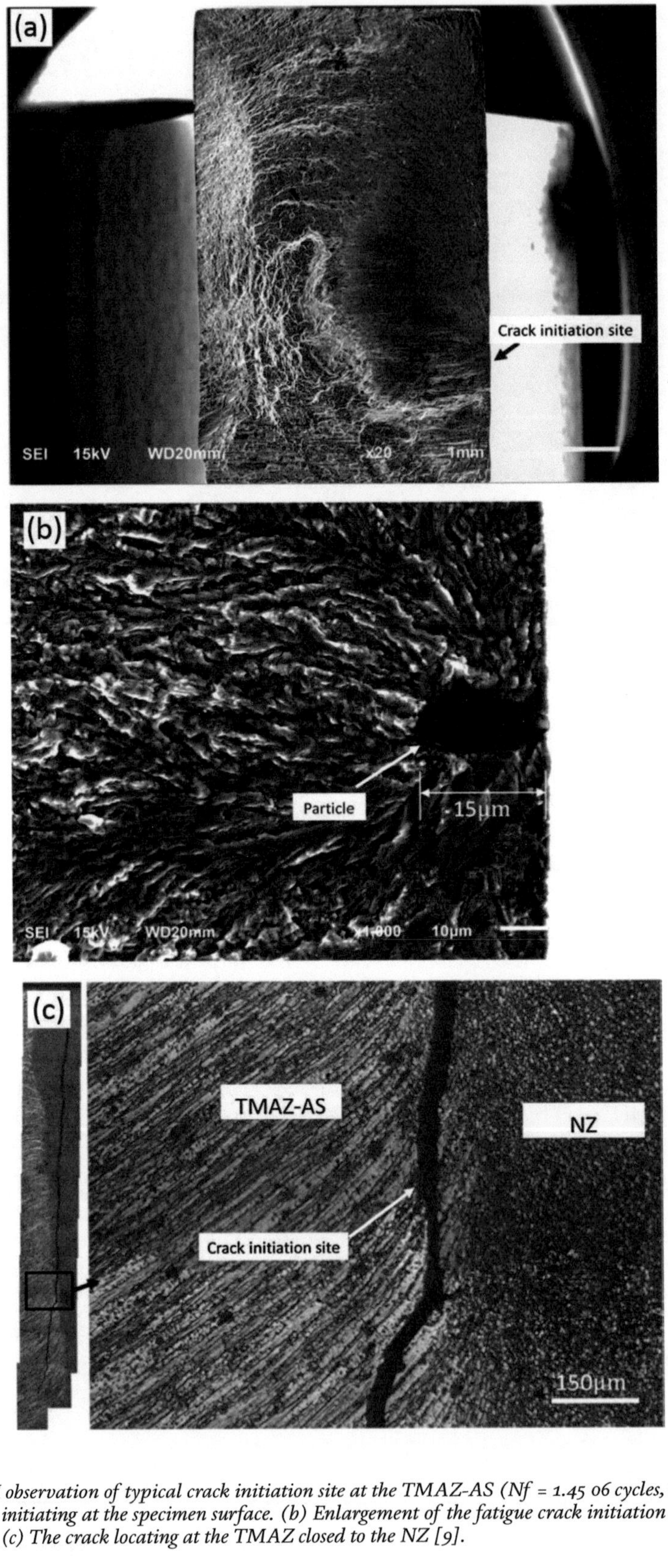

Figure 18.
SEM and OM observation of typical crack initiation site at the TMAZ-AS (Nf = 1.45 06 cycles, rb = 90 MPa). (a) The crack initiating at the specimen surface. (b) Enlargement of the fatigue crack initiation site showing a dark particle. (c) The crack locating at the TMAZ closed to the NZ [9].

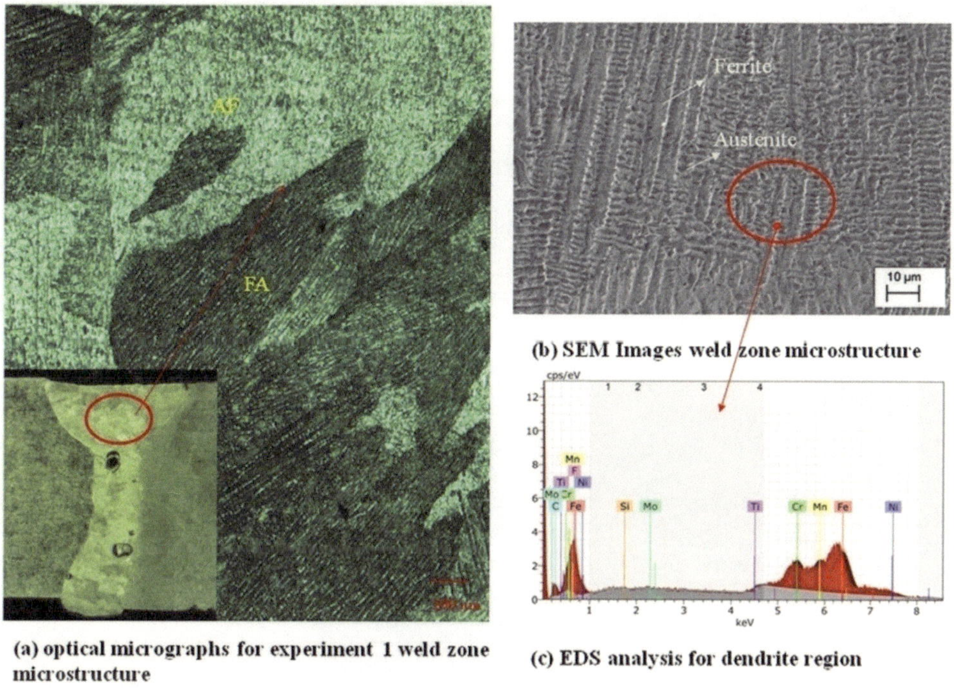

Figure 19.
SEM micrograph of laser welds of steel.

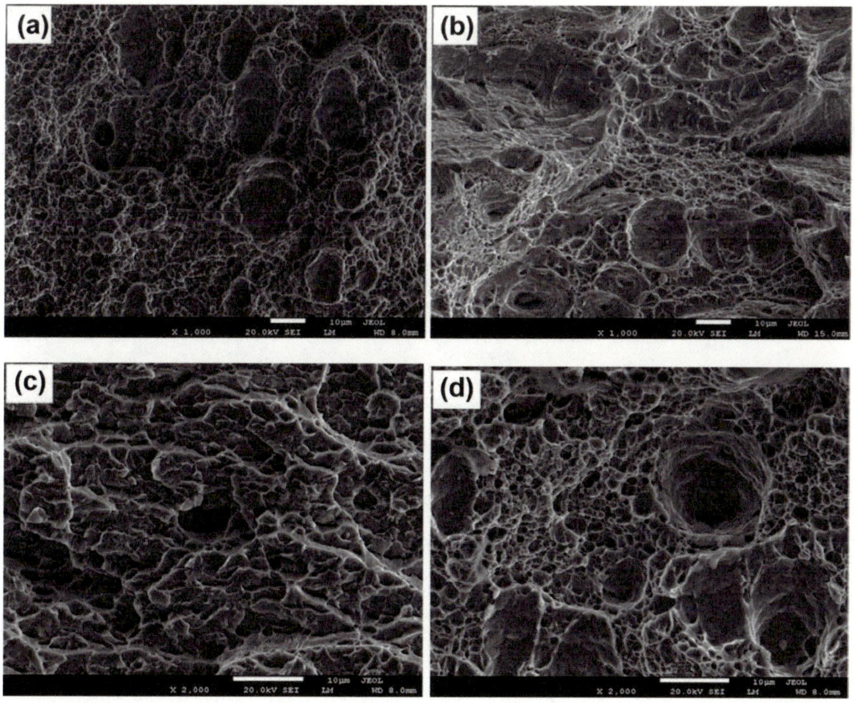

Figure 20.
(a–d) SEM images of tensile facture surface of base metal and welded joints, respectively.

Another case study is the welding of superelastic materials by laser welding. Superelastic NiTi alloy and CuAlMn shape memory alloy have been welded using laser

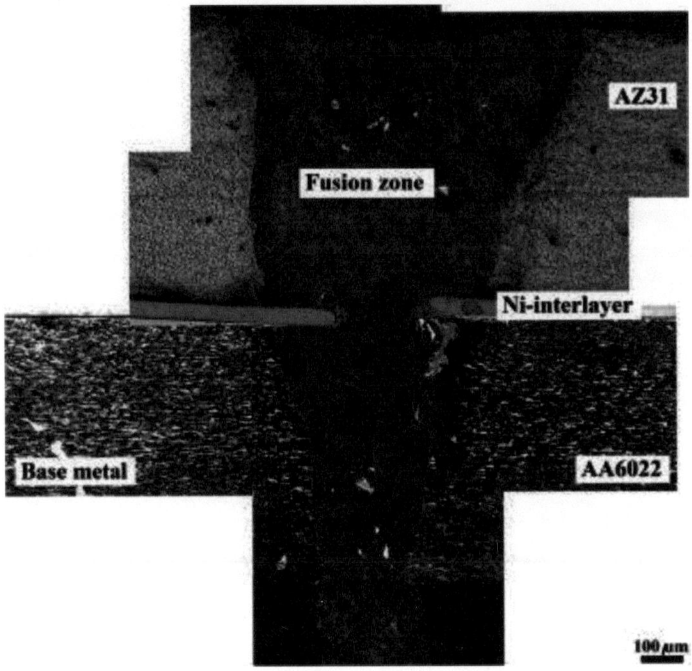

Figure 21.
Laser welds of aluminum-based alloys.

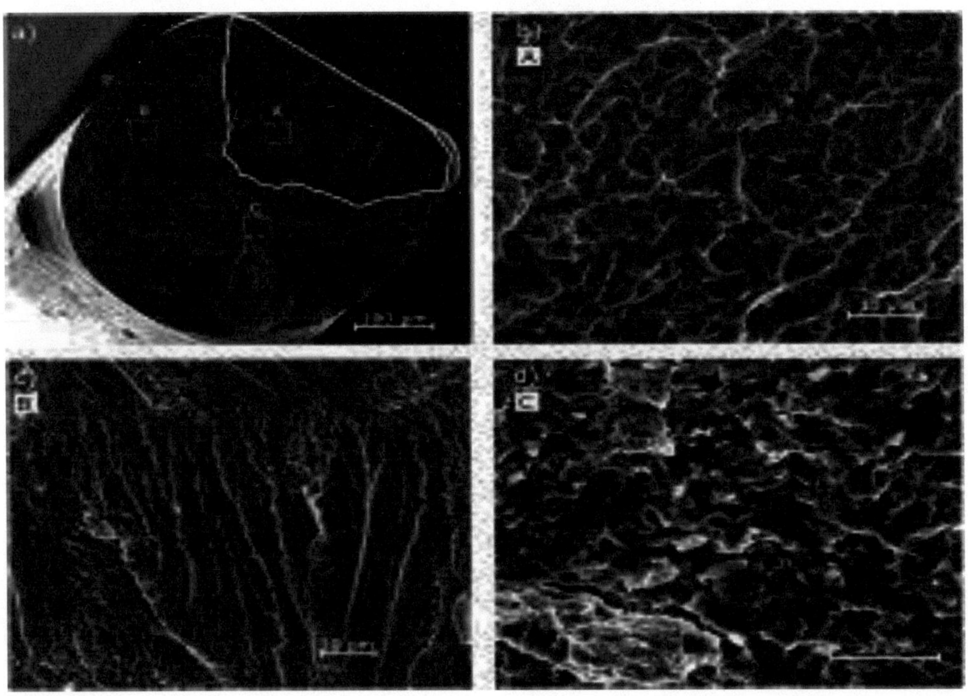

Figure 22.
Laser welded NiTi and CuAlMn [11].

welding. Laser welding has been performed using an Nd-YAG system. The nature of the fractured surface could vary depending upon the system used and the laser welding parameters. For example, when fiber lasers are used, fracture surface may change.

In this case, as shown in **Figure 21**, ductile failure can be seen as also a clear cup and cone type of appearance, which is the characteristic of ductile failure.

Fatigue crack growth behavior can be studied, and SEM micrographs taken in order to get more information about crack growth in dissimilar materials. A comprehensive

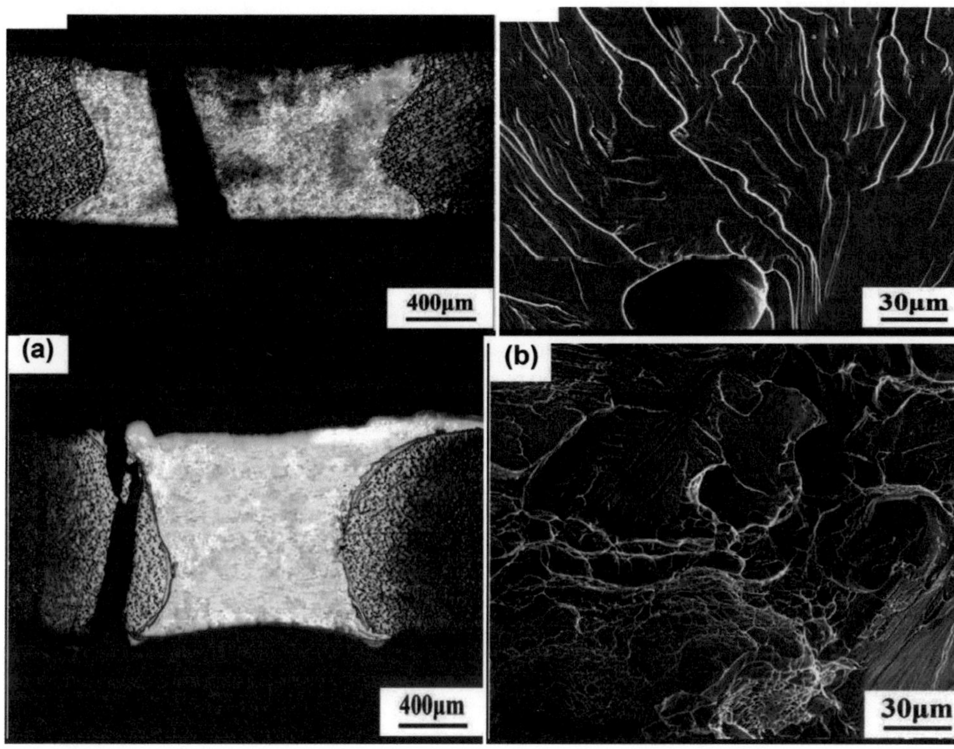

Figure 23.
(a) The fracture path and (b) fracture surface morphology of the joint without filler metal.

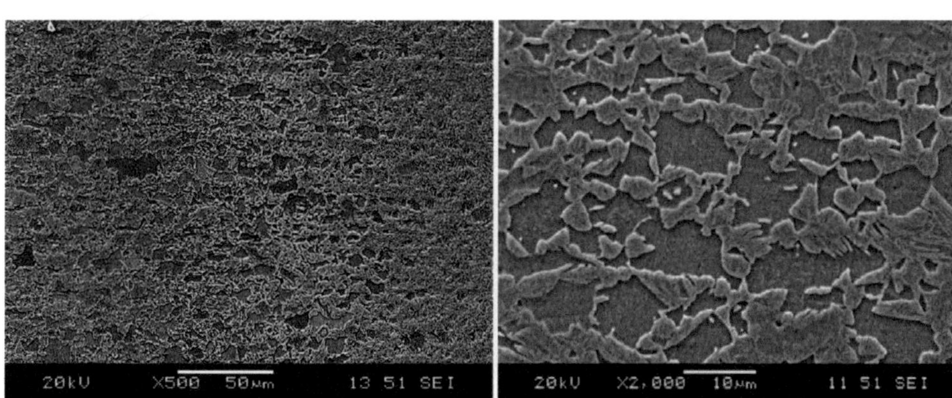

Figure 24.
SEM of fiber laser welds—1 kW fiber laser with 1000 W average output power.

study has been done by Malarvizhi and Balasubramaniam [12]. A good overview of different laser welding processes has been given by Pengfei Wang [13] (**Figure 22**).

Figure 23 show Ti3 Al-Nb laser welds with and without Nb filler. It is seen that Nb filler contributes a lot to ductile behavior as seen below.

Figure 23 shows the fracture path and fracture surface morphology of the joint with Nb filler metal, which is different from those shown in **Figure 12**. The crack was initiated at the weld toe due to generating stress concentration under tensile stress and propagated in HAZ1. The fracture surface also had some cleavage fracture features, but fine dimples and tear-shaped marks along the dimple boundaries were observed (**Figure 13b**). Therefore, the significant enhancement of the joint properties is mainly attributed to the Nb filler metal improving the weld microstructure.

3.5.3 Fiber laser welds

Fiber laser welds show two distinct zones—the intercritical and subcritical HAZ [14–16] (**Figure 24**).

4. Conclusions

This chapter has discussed the scenarios prevailing during fracture of friction welded, friction stir welded, and laser welded joints. A few case studies have been researched upon, both from the author's earlier work and the work of other researchers in the field of solid-state welding. The SEM micrographs' morphologies have been studied, and different fracture surfaces under various conditions of welding have been identified. It is found that the mode of fracture strongly depends upon the welding conditions.

Author details

Dattaguru Ananthapadmanaban[1*] and K. Arun Vasantha Geethan[2]

1 Department of Mechanical Engineering, SSN College of Engineering, Kalavakkam, India

2 Department of Mechanical Engineering, St. Josephs Institute of Technology, Chennai, India

*Address all correspondence to: ananthapadmanaban.dattaguru@gmail.com

References

[1] Zainulabdeen AA, Abbass MK. Investigation of fatigue behaviour and fractography of dissimilar friction stir welded joints of aluminum alloys 7075-T6 and 5052-H34. International Journal of Materials Science and Engineering. December 2014;**2**(2):115-121

[2] Ananthapadmanaban D. PhD thesis. Sathyabama University; 2013

[3] Ravikumar E. PhD thesis. Sathyabama University; July 2019

[4] Shubhavardhana RN, Surendran S. Microstructure and fracture behavior of friction stir lap welding of dissimilar metals. Engineering Solid Mechanics. 2018:7**2019**

[5] Vijayakumar R, Kannan V, Natarajan A. Friction Stir Welding of Aluminium Alloys. IntechOpen; 2017. pp. 81-98

[6] Feng AH, Chen D, Ma ZY. Microstructure and cyclic deformation behaviour of a friction-stir-welded 7075 Al alloy. Metallurgical and Materials Transactions A. April 2010;**41a**:957

[7] Zhang K, Fang Y, Luan G, Zhang J, Hu F. Mechanical and fatigue property of stationary shoulder friction stir welding AA6005. Transactions of the China Welding Institution. 2017;**38**:25-28

[8] He C, Liu Y, Dong J, Wang Q, Wagner D, Bathias C. Fatigue crack initiation behaviors throughout friction stir welded joints in AA7075-T6 in ultrasonic fatigue. International Journal of Fatigue. 2015;**81**:171-178

[9] He C, Liu Y, Dong J, Wang Q, Wagner D, Bathias C. Fatigue crack initiation behaviors throughout friction stir welded joints in AA7075-T6 in ultrasonic fatigue. HAL Archives. January 2018

[10] Oliveira JP, Zeng Z, Andrei C, Braz Fernandes FM, Miranda RM, Ramirez AJ, et al. Dissimilar laser welding of superelastic NiTi and CuAlMn shape memory alloys. Materials & Design. 2017;**128**:166-175

[11] Meng W, Li Z, Huang J, Wu Y, Katayama S. Microstructure and softening of laser-welded 960 MPa grade high strength steel joints. Journal of Materials Engineering and Performance. 2014;**23**:538-544

[12] Malarvizhi S, Balasubramanian V. Fatigue crack growth resistance of gas tungsten arc, electron beam and friction stir welded joints of AA2219 aluminium alloy. Materials and Design. 2011;**32**:1205-1214

[13] Wang P, Chen X, Pan Q, Madigan B, Long J. Laser welding dissimilar materials of aluminum to steel: An overview. The International Journal of Advanced Manufacturing Technology. December 2016;**87**(9-12):3081-3090

[14] Parkes D, Xua W, Westerbaan D, Nayak SS, Zhou Y, Goodwin F, et al. Microstructure and fatigue properties of fiber laser welded dissimilar joints between high strength low alloy and dual-phase steels. Materials and Design. 2013;**51**:665-675

[15] Ravikumar E, Arunkumar N, Anathapadmanaban A, Prabhakaran V. Corrosion Studies on Friction Welded Aluminium Alloy AA6061-T6 to Copper with Nickel Interlayer, Advances in Manufacturing Processes. Lecture Series in Mechanical Engineering. Springer; September 2018

[16] Khodabakhshi F, Shah LH, Gerlich AP. Dissimilar laser welding of an AA6022-AZ31 lap-joint by using Ni-interlayer: Novel beam-wobbling technique, processing parameters, and metallurgical characterization. Optics & Laser Technology. April 2019;**112**(15):349-362

Chapter 6

An Effective Approach for Turbine Hot Component Failure Analysis

Lei-Yong Jiang and Prakash Patnaik

Abstract

One advantage of computational fluid dynamics (CFD) is its ability to reveal the physics or nature of practical engineering problems in detail, allowing engineers and scientists to develop rigorous, effective, and efficient solutions. In this chapter, an effective approach to investigate gas turbine hot component failure is demonstrated, and the mid-span cracking of nozzle guide vanes (NGVs) is used as an example. It is a two-step approach. In the first step, a 60° combustor sector with simplified NGVs and thermocouples attached is simulated; and in the second step, NGV sectors are simulated, where each NGV sector is comprised of one high-fidelity probe NGV and several dummy NGVs. The former identifies the NGV having the highest thermodynamic load and provides the inlet boundary conditions for the latter. The CFD analysis successfully identified the root causes of the NGV damage pattern and mid-span cracking, i.e., the hot streaks from the combustor and inadequate internal cooling.

Keywords: computational fluid dynamics, gas turbine, nozzle guide vane, internal air cooling, failure analysis

1. Introduction

To achieve high thrust, the first-stage nozzle guide vanes (NGVs) of a gas turbine engine are always exposed to high-temperature, high-pressure, high-dynamic load environments and are consequently recognized as one of the most failure-prone components [1]. The thermodynamic load around and inside NGVs varies significantly, which causes considerably uneven structural stresses. For reliable structural, material, and lifetime analysis, a detailed NGV thermodynamic load is deemed necessary.

Due to the harsh conditions, experimental measurements inside the engine are extremely difficult. Computational fluid dynamics (CFD) analysis is routinely used and continuously validated to calculate NGV thermodynamic loads. The heat transfer between the NGV/shrouds and hot-gas flow of a gas turbine combustor was simulated by Mazur et al. [2]. They pointed out that it was feasible to assess the NGV remaining lifetime from the predicted thermodynamic loads and stress and creep strain analyses. Nonetheless, in the paper the shroud thermal boundary conditions were not provided. The heat transfer simulation of a film-cooled linear cascade was performed and validated against the measured data by Harsqama et al. [3]. It was found that the CFD calculations were in good agreement with the measurements of the vane surface Mach number, film cooling row discharge coefficients, wall heat flux and adiabatic film cooling effectiveness. Heidmann et al. [4] developed a conjugate heat transfer solver, which did not require mesh in the solid regions. This code was applied to the heat transfer analyses of the middle-span section of a film-cooled turbine vane for several materials. It was found that the conjugate heat transfer for a film-cooled vane was complicated, and the 1-D analysis between the vane surface and the plenum surface was not suitable.

However, the detailed numerical analysis of air-cooled nozzle guide vanes and shrouds with high-fidelity geometry and well-defined boundary conditions is rare. There are several reasons for this. First, the geometry of real-world turbine NGV/shrouds is complex, particularly for internal cooling flow passages. Second, a fine mesh is needed near NGV walls to calculate properly the heat transfer [5], and thus the mesh size will be large even for a single NGV/shroud assembly. Third, it is difficult to obtain reliable boundary conditions, when both the engine combustor flow field and secondary air cooling flow are involved. The important effect of secondary air-cooling flow on the end walls of an air-film-cooled NGV was pointed out by Charbonnier et al. [6]. Fourth, perhaps the most important reason, the combustor and NGVs are directly connected, and their flow fields are closely coupled. To investigate the NGV failure adequately, the combined combustor-NGV simulation is preferred, particularly for engines where the temperature distribution at the combustor exit or NGV inlet is not uniform. Moreover, to calculate the NGV thermodynamic load with acceptable accuracy, a single high-fidelity NGV simulation is not suitable since the flow passing over each NGV is not the same, and thus an NGV sector simulation is required. The requirement for the combined high-fidelity combustor and NGV sector simulation possibly implies an impractical mesh size, thus the required computing power and time can be enormous [7].

In this chapter, an effective approach to predict the thermodynamic load of engine hot components is presented, and the failure analysis of one type of internally air-cooled NGVs is used as a demonstration case. The approach follows two steps. In the first step, a 60° combustor sector with simplified NGVs and thermocouples is simulated; and in the second step, NGV sectors are simulated where each NGV sector is comprised of one high-fidelity probe NGV and several dummy NGVs. The former identifies the NGV having the highest thermodynamic load and provides the inlet boundary conditions for the latter.

In the following sections, the computational domains and meshes of the combined combustor-NGVs and NGV sectors, numerical methods, engine operating conditions, boundary conditions, and results at flight conditions are presented and discussed. Finally, the chapter provides conclusions and recommendations going forward.

2. High-fidelity CFD models

2.1 Computational domain and mesh of combined combustor-NGVs

The computational domain and mesh, including a 60° sector of a traditional can-annular gas turbine combustor and simplified 10 NGVs and 3 thermocouples, is shown in **Figure 1**. The whole mesh is shown in **Figure 1(a)**; the close-up view of 10 NGVs, 3 thermocouples, and 10 NGV cooling air supply slots at the end of the annular chamber is presented in **Figure 1(b)**; and the high-fidelity combustion can mesh is illustrated in **Figure 1(c)**. In **Figure 1(c)**, the meshed components include the fuel injector, dome swirler, wiggle strips, baffle and baffle splashing holes (underneath baffles), and primary and secondary dilution holes. The external surfaces of the simplified NGVs were meshed, and the NGV internal cooling flow passages and solid metal regions were not meshed in this step. To comprise the cooling effect, the

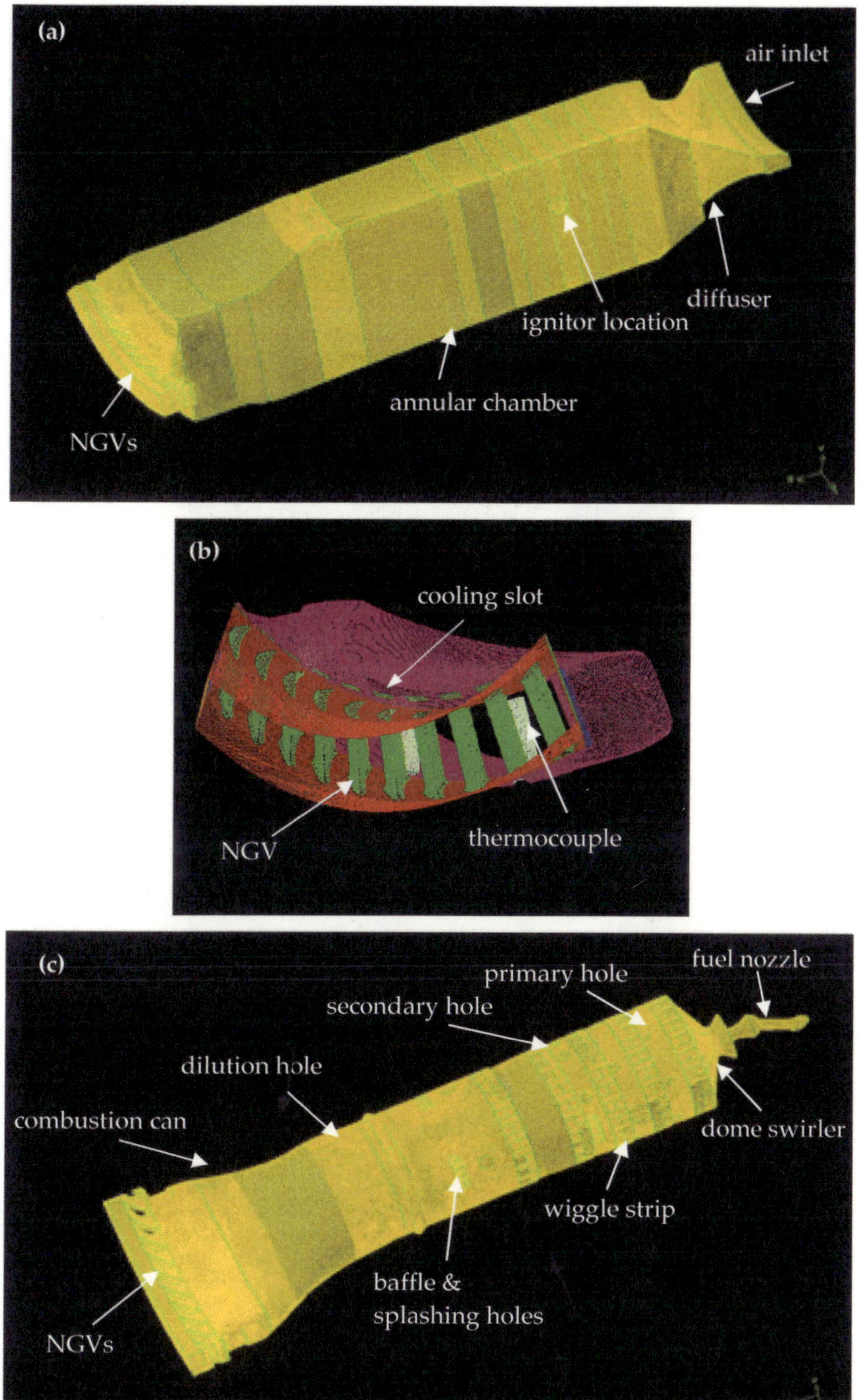

Figure 1.
Computational domain and mesh of the combined combustor-NGVs: (a) the whole mesh, (b) the NGV and thermocouple surface mesh, and (c) the can, NGVs, and thermocouple mesh.

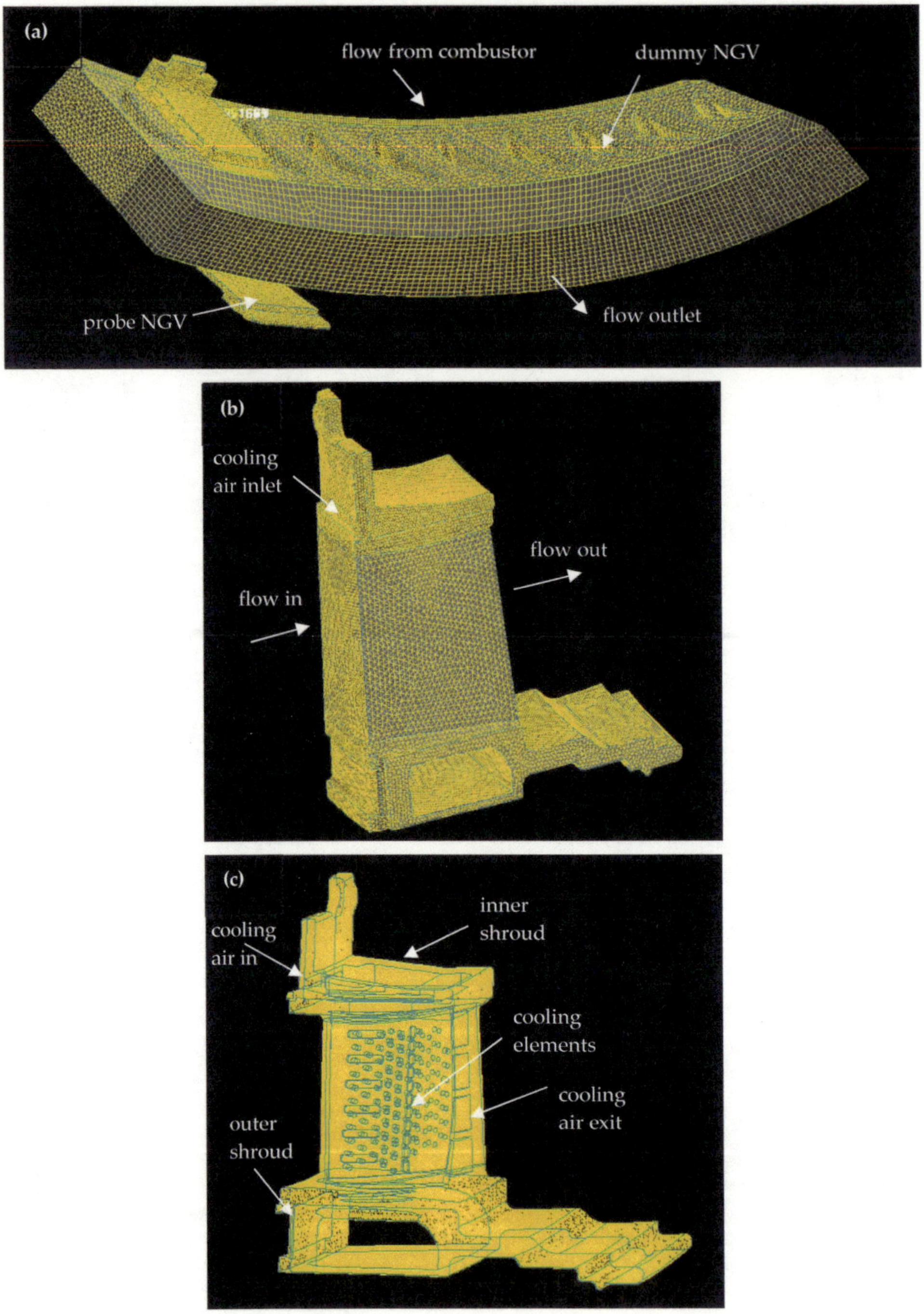

Figure 2.
NGV-2 sector domain and mesh: (a) the whole mesh, (b) the probe NGV mesh, and (c) the probe mesh in metal regions.

mean heat flux obtained from a previous single NGV simulation [8] was defined at the NGV external surfaces. More information regarding the mesh and mesh independence issue is described in Jiang and Corber [9], and the mesh size used in the simulation was 17.4 million.

As illustrated in the above figure, compressed air flows into the annular chamber throughout a narrow annulus and then flows over and enters the combustion can through air management elements around the can or liner. In the combustion can, fine fuel droplets generated by an air-assist fuel nozzle vaporize, mix with air, and then burn, and finally the combustion products arrive at the thermocouples and NGVs.

2.2 Computational domain and mesh of NGV sector

A 60° NGV sector mesh with one high-fidelity NGV (probe NGV) and nine dummies is shown in **Figure 2(a)**, where the probe NGV is located at the second position from the left-hand side, looking upstream. The probe NGV meshes for the main portion of its domain and for regions constructed from metal are given in **Figure 2(b)** and **(c)**, respectively.

The computational domain of the NGV sector starts at the middle cross section between the NGVs and thermocouples and extends to half NGV chord length in the downstream direction. To properly solve the conjugate heat transfer, fine nodes were prescribed near the probe NGV inner and outer walls, and cooling air slot, holes and elements; and coarse nodes were generated in the regions away from the probe NGV walls. Coarse meshes were also used for the dummy NGVs. More information regarding the probe mesh and y+ value is available in [8]. A total of 7.8 million cells were used for the NGV sector simulations, and 6.6 million cells among them were for the probe NGV.

There are 10 NGVs behind the combustion can, and as seen in Section 4.1, the NGV at the second position, NGV-2, is subjected to the highest thermal load, and the NGV at the ninth position, NGV-9, experiences the lowest thermal load. For the purpose of comparison, the NGV-9 sector mesh similar to that shown in **Figure 2** was also created.

Figure 2 illustrates that the combustion gas mixture from the combustor passes through the NGV sector domain. The cooling air coming from the end of the annular chamber as shown in **Figure 1** flows into the probe NGV body at the cooling air inlet of the inner shroud, enters the internal cooling flow passages, passes through the four rectangular opens, then merges with the main flow outside the NGV, and finally runs out of the sector domain. In order to enhance the heat transfer among cooling air and NGV body, many cooling elements are constructed on either side of the cooling chamber inside the NGV though they are not clearly seen here. The cooling air absorbs heat from the NGV body through impingement and convection heat transfer and then reduces the NGV metal temperature.

3. Numerical methods

3.1 Numerical approach and physical models

For the first step, the standard Eulerian-Lagrangian method was used to solve the two-phase, steady, turbulent, compressible, reacting flows; and in the second step, steady, turbulent, compressible flows were considered [10, 11]. In both cases, the Favre-averaged governing equations for mass, momentum, species, and total enthalpy were resolved. For the coupled or conjugate heat transfer between the flow fields and solid metal regions, both fluid and metal regions were computed simultaneously. In the metal regions, since only heat conduction was concerned, the energy governing equation became simpler than that for fluid fields.

To close the flow-governing equations, the turbulence transfer terms and species and energy sources had to be properly modeled. For the combined combustor-NGV simulations, the eddy dissipation (EDS) combustion model and discrete ordinate radiation model were used to account for species and energy source terms, and the realizable k-ε model was selected for turbulence momentum and scalar transfers. These models were carefully evaluated with the comprehensive database measured from a model combustor [12–14]. For the air-splitting over the air management elements of the combustion can, the predicted values were closely correlated to the results calculated from the semi-empirical correlations of discharge coefficients [9]. For the NGV sector simulations, the shear stress transport (SST) k-ω model was selected since it demonstrated substantial improvements in the simulations of adverse pressure gradient and separation flows in comparison with the standard k-ε and k-ω models [15]. It was successfully applied to the heat transfer simulation of a film-cooled linear cascade, in good agreement with the measured experimental results [6].

3.2 Engine operating and boundary conditions

For the combined combustor-NGV simulation, the engine takeoff conditions from reference [16] were considered in the present analysis. The air and fuel flowrates, pressure, and temperature were defined as boundary conditions at the combustor inlet for the combined combustor-NGV case. To provide the turbulent kinetic energy and dissipation rate at the air inlet, the hydraulic diameter of the inlet annulus and 5% turbulence intensity were used. At the side surfaces of the 60° sector, periodic boundaries were specified. The size and velocity components of fuel droplets in the nozzle radial direction were measured at the National Research Council of Canada's High-Pressure Spray Facility with a phase Doppler particle analyzer [9], and these data were used to define the fuel spray from the nozzle. The fuel mass flux data in the nozzle radial direction were from Rizk et al. [17]. They carried out the measurements with a mechanical patternator in a high-pressure rig at Purdue University, Indiana. Fourteen fuel spray cones (1800 droplets each) were specified, and a total of droplets in the simulation were 25,200. To consider the effect of flow tangential velocity on the pressure distribution, a radial equilibrium pressure distribution was enforced at the computational domain exit.

As mentioned earlier, the simulation results from the combined combustor-NGV simulation were used as the inlet boundary conditions for the NGV sector simulations, including the total pressure, total temperature, velocity direction, turbulence kinetic energy, and turbulence dissipation rate at the mid-cross section between the NGVs and thermocouples. To obtain the specific dissipation rate, ω, for the SST turbulence model at the inlet, the following equation was used:

$$\omega = \varepsilon / (0.09\,k) \tag{1}$$

where ε and k are dissipation rate and turbulence kinetic energy, respectively [18]. The NGV sector domain exit static pressure was defined from the combined combustor-NGV simulation, and it was slightly adjusted to maintain the mass flow rate. For the NGV internal cooling, 2.5% of the total inlet airflow was used. At the domain exit, the radial equilibrium pressure distribution was also enforced, and periodic boundaries were specified at the side surfaces of the 60° NGV sector. Based on the information of Mach number and flow temperature of the secondary airflow from Snedden et al. [19], the thermal boundaries on the external surfaces of shrouds were estimated, and please refer to [8] for details.

To have an overview of the NGV heat transfer processes, the Biot number of the NGV was calculated by the following equation:

$$Bi = \frac{L_c h}{k} \tag{2}$$

where h is the mean heat transfer coefficient over the NGV airfoil external surface, L_c is the NGV airfoil characteristic length, defined as the volume divided by the surface area, and k stands for the thermal conductivity of the airfoil body. The NGV is made of the X-40 alloy [20] with the density of 8610 kg/m^3, specific heat of 411 J/kg-K thermal, and conductivity of 22.8 W/m-K evaluated at 900 K. The Biot number for the NGV-2 is 0.03 which is much smaller than 0.1. It implies that the heat transfer resistance of conduction in the NGV metal body is significantly less than that of convection through the NGV/flow interfaces [21].

3.3 Solution methods

ANSYS CFD Premium, a commercial code, was used in all simulations [11]. A pressure-based coupled solver with a second-order accurate scheme was used to solve the flow fields. All simulations were well converged. For velocity components and scalar items, the normalized residuals were less than 4×10^{-5}, and for turbulent variables, they were about 6×10^{-4}. The monitored flow variables stayed unchanged for the first four digits, and it indicated that the flow steady condition is reached.

4. Results and discussion

4.1 Results of combined combustor-NGV simulation

The combined combustor-NGV simulation has provided a large amount of information. The detailed flow features have been described in [9], including the major flow parameter distributions and complex vortex structures across the dome swirler, primary and secondary dilution holes, wiggle strips, baffles, and baffle splashing holes, as well as along the mid-longitudinal plane. To give an overall picture, a few flow parameter contours along the longitudinal plane are presented here. Moreover, as required for the NGV sector simulations, the flow parameter distributions at the middle cross section between NGVs and thermocouples are discussed.

The contours of normalized static temperature, Mach number, fuel mass fraction, and normalized static pressure at the combustion can mid-longitudinal section are displayed in **Figure 3**, where the temperature and pressure are normalized by their maximum values, respectively. As seen in **Figure 3(a)**, the temperature is high in the upstream region and relatively low in the downstream region. **Figure 3(b)** indicates that Mach number is low in most of the can, increases gradually in the contraction section, and quickly reaches a maximum in the NGV section. As shown in **Figure 3(c)**, the fuel, C12H23, is mainly located in the one third upstream region of the can, which implies the chemical reaction mainly occurs in this upstream region. **Figure 3(d)** reveals that the pressure remains almost constant in most of the can region and decreases quickly in the NGV section due to flow acceleration.

The zoomed views of **Figure 3(b)** and **(d)** near the NGV section are illustrated in **Figure 4(a)** and **(b)**, where the mid-longitudinal plane cuts through one NGV and one thermocouple. Note that the NGV and thermocouple are outlined by thin white lines. At the thermocouple stagnation point and in the wake flow region behind the thermocouple, as anticipated, the Mach number is low; and at the thermocouple stagnation point, the pressure is somewhat higher than that in the wake flow. The typical NGV flow features are observed in these two plots: the Mach number is high on the suction side and low on the pressure side; and the pressure is low on the suction side due to flow acceleration and high on the pressure side.

The contours of total temperature, total pressure, turbulent kinetic energy, and turbulence dissipation rate at the middle section between the NGVs and thermocouples are shown in **Figure 5**. In these plots the flow variables are normalized

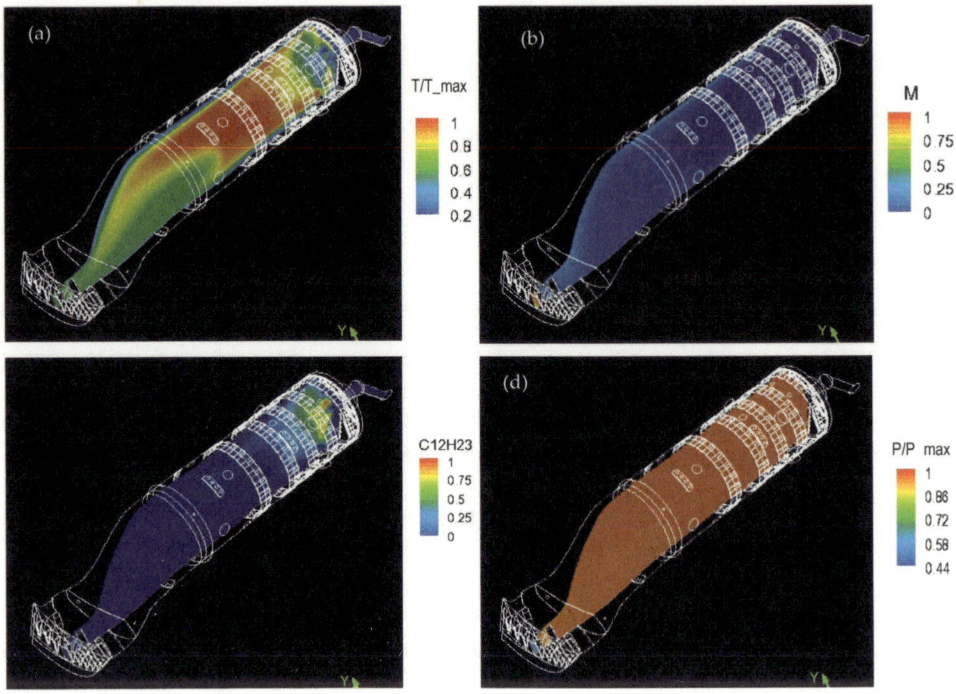

Figure 3.
Flow parameter distributions at the mid-longitudinal section: (a) normalized static temperature, (b) Mach number, (c) C12H23 mass fraction, and (d) normalized static pressure.

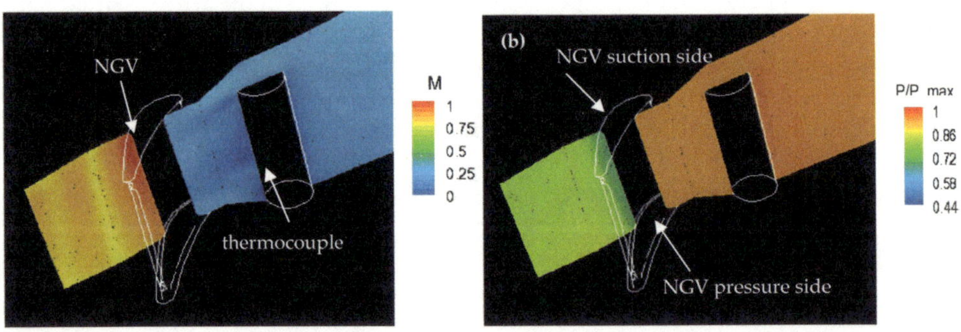

Figure 4.
(a) The zoomed view of **Figure 3(b)** *near NGV section and (b) the zoomed view of* **Figure 3(d)** *near the NGV section.*

by their section maximum values, respectively. **Figure 5(a)** clearly shows the hot streaks or temperature distortions from the combustor. The normalized temperature changes from 0.61 to 1.0 and has an average value of 0.888. These values are in excellent agreement with the experimental data, $T\_average = 0.886$ and $T\_max = 1.0025$ [17]. The deviations in the mean and maximum temperatures are less than 0.5%. This provides more confidence for the subsequent NGV failure analysis.

The cause of the hot streaks from the combustion can exit can be readily clarified by its geometry. As shown in **Figure 6**, the largest and second largest dilution holes just upstream of the contraction section are asymmetrically located with reference to the can mid-longitudinal plane, and this is the main reason of the temperature distortion at the combustor exit.

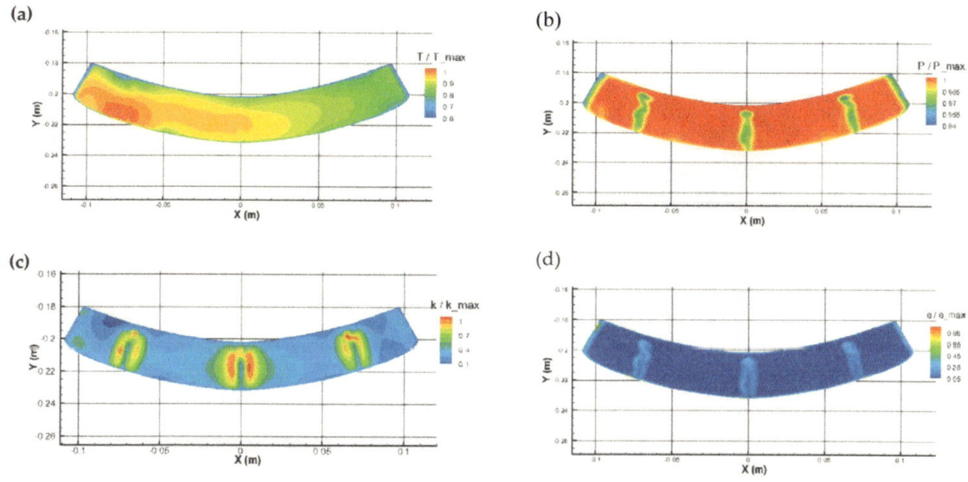

Figure 5.
Normalized flow parameter contours at the cross section between the NGVs and thermocouples: (a) the total temperature, (b) the total pressure, (c) the turbulent kinetic energy, and (d) the turbulence dissipation rate.

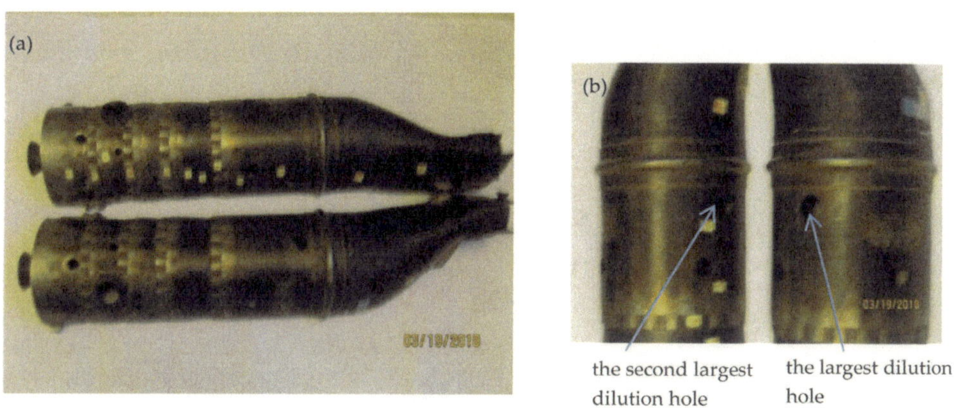

Figure 6.
Combustion can: (a) two halves and (b) the two dilution holes before the contraction section.

As observed in **Figure 5(b)**, there are three narrow relatively low total pressure regions at the cross section, which are the wake flows of the three thermocouples. Moreover, the total pressure near the two-side boundaries is also low. This is mainly due to the flow passage increase in the circumferential direction to house two-side NGVs. **Figure 5(c)** and **(d)** reveals that the turbulent kinetic energy and dissipation rate are high in the wake flow regions of the thermocouples. It is interesting to note the rotating vortices in these local regions, from near the can inner wall to the can outer wall. It is believed that this phenomenon is caused by the gaps between the thermocouples and the can inner wall (please see **Figure 1(b)** or **Figure 4**). As usual, the turbulence dissipation rate near the surrounding walls is high.

Figure 7 displays the temperature distributions at the outer surfaces of the simplified 10 NGVs and at the mid-cross section between the NGVs and thermocouples. The comparison of temperature variations and mean temperatures over these NGVs indicates that among the 10 NGVs, the highest surface maximum temperature and mean temperature occur at NGV-2, while the lowest maximum and mean values happen at NGV-9. These results indicate that NGV-2 is the most vulnerable NGV under the flight condition, and this agrees with the field-observed NGV damage pattern.

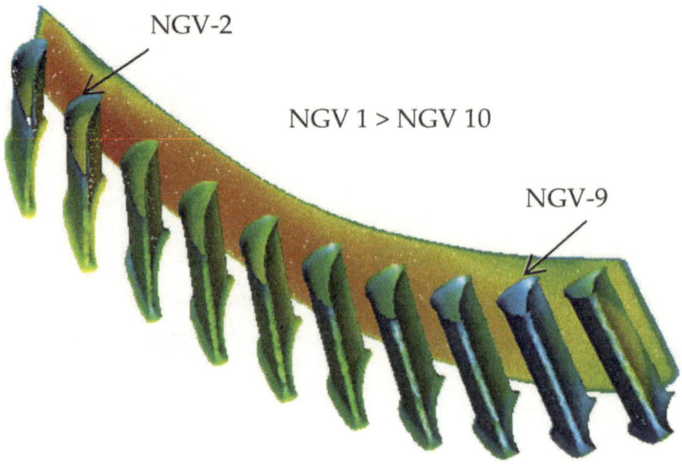

Figure 7.
Normalized static temperature contours at the external surfaces of the simplified NGVs and at the middle cross section between the NGVs and thermocouples.

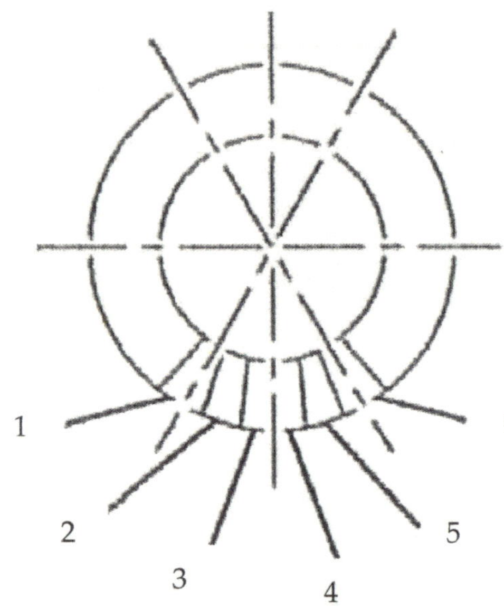

Figure 8.
The locations of six NGV units relative to the combustion can exit, looking upstream.

The locations of six NGV units relative to the combustion can exit are outlined in **Figure 8**. A 60° sector is formed by two dashed lines, and the overlapped line with the small circle defines the combustion can exit. Each NGV unit has two NGVs, and the leading NGV and trailing NGV are named counterclockwise, looking upstream. **Figure 8** indicates that the trailing NGV of Unit 1 and the leading NGV of Unit 6 are within the can exit domain. Thus there are 10 NGVs behind each combustion can. Most of the failures are Unit 2, while a few are Unit 1 or Unit 6, the clockwise neighbors of Unit 2.

4.2 Results of NGV sector simulations

A large amount of data was obtained from the NGV sector simulations at the flight conditions. The main features of the complex flow field and heat transfer have been previously described in [8]. Here the results related to the NGV failure analysis are presented and discussed.

Shown in **Figures 9** and **10** are the external surface temperatures of the NGV/shrouds for the three cases: NGV-2, NGV-2 with the averaged inlet flow parameters, and NGV-9, where the temperatures are normalized by the same maximum values.

Figures 9–11 clearly reveal that the thermal load of NGV-2 is highest among the three cases. Its maximum surface normalized temperature reaches 0.96, 6.8% higher than NGV-2 with the mean inlet conditions, and 10.7% higher than NGV-9. The mean surface temperature of NGV-2 is 0.80, 2.5% higher than the second case and 5.7% higher than the third case. These results confirm the strong hot streak effect on the NGV thermal load, where a large increase in the metal temperature can result in significant decrease in strength [22]. Some means of thermal protection, such as the addition of thermal barrier coating (TBC) and/or improvement of the internal cooling arrangement, should be considered for the vulnerable NGVs.

Of greater significance in **Figures 9–11** is that for all three cases, there is a nearly circular high-temperature zone in the middle of the NGV pressure side, shifted slightly towards the leading edge and outer shroud. In contrast, on the suction side, there is no high-temperature zone observed in the middle, and the temperature variation is less than for the pressure side. Note that for the second case, the hot streak effect is not included since the averaged inlet boundary conditions are used. These observations clearly explain the root cause of the mid-span cracking of the NGVs,

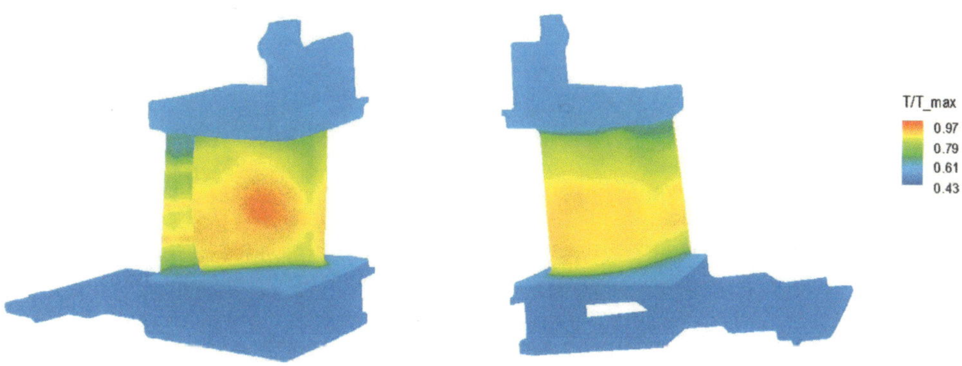

Figure 9.
Normalized temperature contours at the external surfaces of the NGV/shrouds for the probe NGV at the second position, NGV-2.

Figure 10.
Normalized temperature contours at the external surfaces of the NGV/shrouds for the probe NGV at the second position, NGV-2, with the averaged inlet flow conditions.

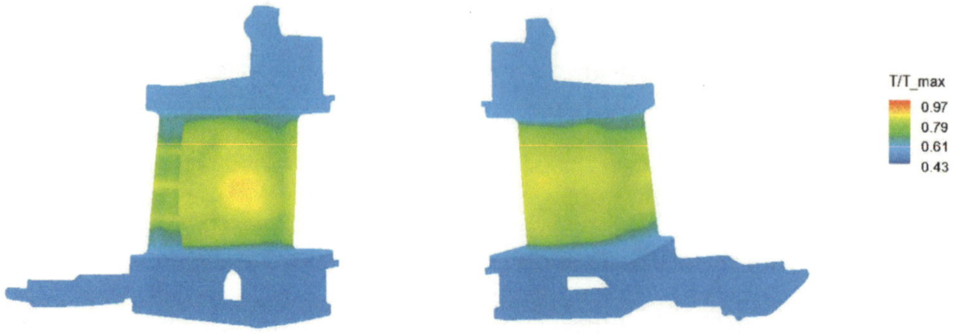

Figure 11.
Normalized temperature contours at the external surfaces of the NGV/shrouds for the probe NGV at the ninth position, NGV-9.

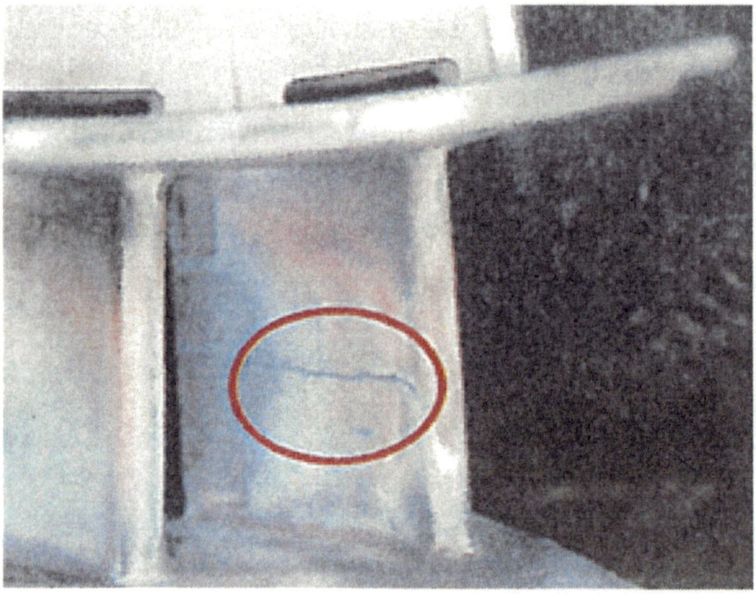

Figure 12.
Typical mid-span cracking on the pressure sides.

which are consistent with the field-observed NGV damages as shown in **Figure 12**. As shown in in Figure 12, the crack is at the mid-span of the NGV pressure side.

For the above three cases, **Figures 13–15** present the distributions of static temperature, Mach number, and velocity vectors at the section across the 5th trailing-edge hole and the 11th leading-edge cooling hole counted from the cooling air inlet. The maximum incoming flow temperature is used to normalize temperature contour plots. The maximum incoming temperature is 1.0 for the first case, 0.91 for the second, and 0.87 for the third. The incoming temperature difference between NGV-2 and NGV-9 is 0.13, which results in the considerable difference in their thermal loads. The typical NGV flow features are illustrated in these figures. The temperature is high at the pressure side and gradually decreases along the suction side; and the Mach number is low near the leading edge, gradually increases along the pressure side and increases and then decreases along the suction side. The velocity vectors from the leading-edge cooling hole illustrate the impingement cooling, and the velocity vectors along the internal and external walls represent the convection cooling.

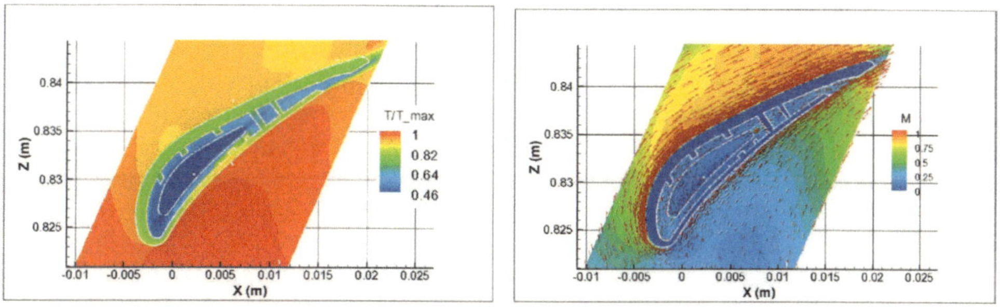

Figure 13.
Normalized temperature and Mach number contours and velocity vectors at the section across one trailing-edge cooling hole and one leading-edge cooling hole for the pin-fin NGV-2.

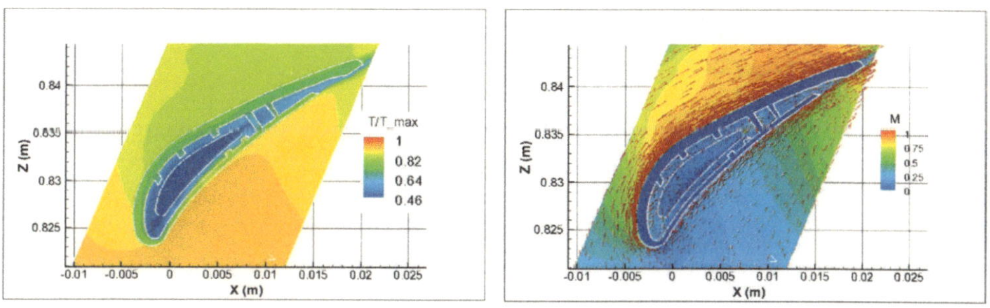

Figure 14.
Normalized temperature and Mach number contours and velocity vectors at the section across one trailing-edge cooling hole and one leading-edge cooling hole for the pin-fin NGV-2 with the averaged inlet flow parameters.

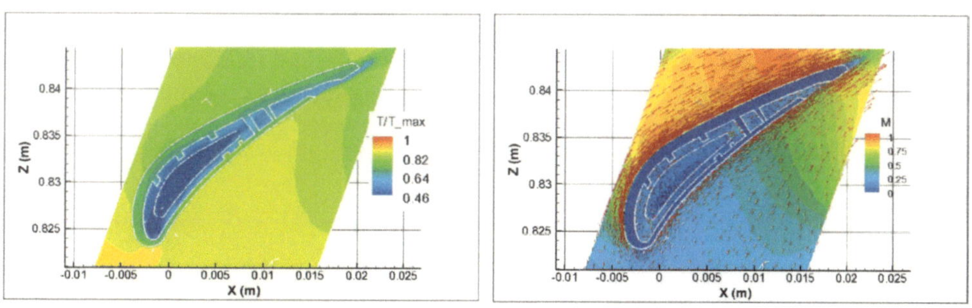

Figure 15.
Temperature and Mach number contours and velocity vectors at the section across one trailing-edge cooling hole and one leading-edge cooling hole for the pin-fin NGV-9.

5. Conclusions

In this chapter, the flow fields and heat transfer of the combined combustor-NGV simulation and the NGV sector simulations have been successfully simulated. The predicted temperature distribution at the combustor exit is in excellent agreement with the experimental data, which provides more confidence in the NGV failure analysis.

The results of the first step analysis indicate that NGV-2 has the highest thermal load and NGV-9 has the lowest thermal load at the flight conditions, which is consistent with the NGV damage pattern. Therefore, it is concluded that the field-observed NGV damage pattern is caused by the hot streaks from the combustor, and this provides a good foundation for the second step analysis.

In the second step analysis, it is found that there is a circular high-temperature zone in the middle of the NGV on the pressure side, shifted slightly towards the leading edge and outer shroud for all cases studied, which is consistent with the location where the mid-span cracking occurs. Consequently, it is concluded that the root cause of the NGV mid-span cracking is the high thermal loading, or stated another way, an inadequate cooling arrangement.

In summary, the CFD analysis has successfully identified not only the root cause of the NGV failure pattern but also the root cause of the NGV failure location. This excellent example demonstrates the effectiveness of the two-step and dummy component approach to investigate turbine hot component failures, and its advantages are as follows:

1. With the dummy component approach, the computing time is greatly reduced. In this example, with nine dummy NGVs, the computing time of the NGV sector simulation is reduced to about one tenth of that without the dummy component approach.

2. The separation of high-fidelity combustor and high-fidelity component sector simulations makes it more practical to obtain the reliable inlet boundary conditions for component sector simulations with accepted computing time.

3. This approach is more suitable to the workplace where the computing power and memory size are limited.

Acknowledgements

The authors are grateful to the Department of National Defense (Canada), as well as Aerospace Research Center, and the National Research Council of Canada for funding and supporting this collaborative research project. Special thanks to Dr. Steve Zan for the valuable comments and suggestions in the preparation of the manuscript.

Author details

Lei-Yong Jiang* and Prakash Patnaik
Aerospace Research Centre, National Research Council of Canada, Ottawa, Ontario, Canada

*Address all correspondence to: lei-yong.jiang@nrc-cnrc.gc.ca

References

[1] Saravanamuttoo HIH, Rgers CFC, Cohen H. Gas Turbine Theory. 5th ed. UK: Pearson Education Limited; 2001

[2] Mazur Z, Hernandez-Rossette A, Garcia-Illescas R, Luna-Ramirez A. Analysis of conjugate heat transfer of a gas turbine first stage nozzle. Applied Thermal Engineering. 2006;**26**:1796-1806

[3] Harsqama SP, Burton CD, Chana KS. Measurements and computations of external heat transfer and film cooing in turbines. In: The Proceedings of the 10th International Symposium on Air Breathe Engines (ISABE); 1991. pp. 1276-1284

[4] Heidmann JD, Kassab AJ, Divo EA, Franklin Rodriguez F, Steinthorsson E. Conjugate Heat Transfer Effects on a Realistic Film-cooled Turbine Vane, ASME IGTI paper, GT2003-38553; 2003

[5] Jiang LY, Manipurath S, Bourque G, Houde M. Flow-field of a triple-walled gas-sampling probe with sub-cooled boiling effect. Flow Measurement and Instrumentation. 2007;**18**:156-116

[6] Charbonnier D, Ott P, Jonsson M, Kobke Th, Cottier F. Comparison of Numerical Investigations with Measured Heat Transfer Performance of a Film Cooled Turbine Vane, ASME IGTI paper, GT2008-50623; 2008

[7] Mongia HC. Recent Advances in the Development of Combustor Design Tools, AIAA Paper 2003-4495; 2003

[8] Jiang LY, Wu X, Zhong Zhang Z. Conjugate heat transfer of an internally air-cooled nozzle guide vane and shrouds. Advances in Mechanical Engineering. 2014;**2014**:146523. DOI: 10.1155/2014/146523

[9] Jiang LY, Corber PA. Air Distribution over a Combustor Liner, IGTI paper, GT-2014-25405; 2014

[10] Stiesch G. Modelling Engine Spray and Combustion Processes. New York: Springer; 2003

[11] ANSYS Fluent Inc. Fluent 19 documentation, 10 Cavendish Court, Lebanon, NH 03766, USA; 2018

[12] Jiang LY, Campbell I. Radiation benchmarking in a model combustor. Journal of Engineering for Gas Turbines and Power. 2009;**131**:011501

[13] Jiang LY, Campbell I. Application of three combustion models to a model combustor. Canadian Aeronautics and Space Journal. 2005;**51**(1):1-11

[14] Jiang, L.Y., Chapter 7: RANS modelling of turbulence in combustors, in The Book of Turbulence Modelling Approaches—Current State, Development Prospects, Applications, Edited by Dr. Konstantin V., IntechOpen, ISBN: 978-953-51-5311-5, 2017

[15] Menter FR. Two-equation Eddy-viscosity turbulence models for engineering applications. AIAA Journal. 1994;**32**(8):1598-1605

[16] Oechsle VL, Ross PT, Mongia HC. High Density Fuel Effects on Gas Turbine Engines, AIAA-87-1829; 1987

[17] Rizk NK, Oechsle VL, Ross PT, Mongia HC. High Density Fuel Effects, Technical Report AFWAL-TR-88-2046, Wright-Patterson Air Force Base, Aero Propulsion Laboratory, USA; 1988

[18] Wilcox DC. Turbulence Modelling for CFD. second ed. La Canada, California, USA: DCW Industries Inc.; 2002

[19] Snedden G, Roos T, Naidoo K. Detailed Disc Assembly Temperature Prediction: Comparison between CFD and Simplified Engineering Methods, ISABE paper 2005-1130; 2005

[20] Brown WF Jr, Setlak SJ. Aerospace Structural Materials Handbook. 38th ed. Purdue University: CINDA/USAF CRDA Handbooks Operation; 2004

[21] Incropera FP, DeWitt DP. Fundamentals of Heat and Mass Transfer. fifth ed. USA: John Wiley and Sons, Inc.; 2002

[22] Klopp WD. Non-Ferrous Alloys, Co. MAR-M-509, Aerospace Structural Metals Handbook; 1985

Chapter 7

A Wear Analysis Carried On Connecting Rod Bearings From Internal Combustion Engines

Juan Rodrigo Laguna-Camacho, Silvia M. Sánchez-Yáñez, Gabriel Juárez-Morales, Maria I. Cruz-Orduña, Luz M. Ramos-González, Cristóbal Cortez-Domínguez, Roberto L. Cabrera-Santiago and Javier Calderón-Sánchez

Abstract

In the present work, an analysis was carried out to know the wear modes present in connecting rod bearings from internal combustion engines. These mechanical elements were selected since they are exposed to different engineering failures such as incorrect assembly, severe loads, extreme temperatures, inadequate conditions, and loss of lubricity. In this particular case, the bearings that were selected had a service life of approximately 8 years. Different techniques such as SEM and optical microscopy, EDS analysis, hardness testing, and surface profilometry were used to characterize the unworn and worn bearings. Wear mechanisms such as sliding wear (scoring), fatigue wear with cracks where torn out material was clearly observed, discoloration areas, and two- and three-body abrasion wear (rubbing marks) were identified on the bearing surfaces.

Keywords: wear modes, sliding wear, fatigue wear, cracks, discoloration, connecting rod bearings

1. Introduction

A connecting rod is a mechanical element that, subjected to tensile or compressive stresses, transmits the articulated movement to other parts of the machine. In an internal combustion engine, this connects the piston to the crankshaft. The rod consists of a rigid rod designed to establish articulated joints at its ends. It allows the union of two operators transforming the rotary movement of one (crankshaft) in the reciprocating motion of the other (piston) or vice versa. Its cross section or profile can have the shape of H, I, or +. This mechanical element can be manufactured by steel, titanium, or aluminum alloy. In the automotive industry, these are produced by forging, but some companies can manufacture by machining.

This mechanical element is used in a multitude of machines that require the conversion between alternative continuous and linear rotary movement. Clear examples of some machines where they are used are steam engine, internal combustion engines, machines moved by the foot, water pumps, etc. The main stresses suffered by the connecting rod are compound bending at the moment of the maximum

 IntechOpen

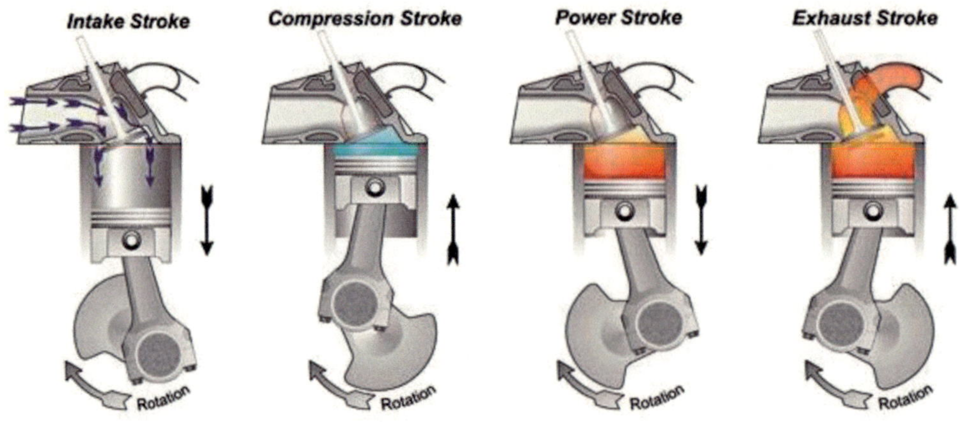

Figure 1.
Internal combustion engine [1].

load when exploiting the fuel mixture (expansion cycle) (third time in **Figure 1**). The compression would be given by the force component on the longitudinal axis of the connecting rod, the bending by the component transverse to it, and the same with the reactive torque provided by the load through the crankshaft when opposing the movement [1]. In addition, the rod undergoes a compressive stress again in the compression stage of the mixture. Due to this, two points are critical; they occur in different stages of the mechanical cycle; the first one is seen during the compression; this takes place in the middle part of the connecting rod; the second critical point is located in the lower part of the rod and occurs during the cycle expansion. The screws, on the other hand, support only a small percentage of the load.

In this work, an analysis of the wear that occurs in the connecting rod bearings from internal combustion engines was performed. It is due to the high stresses to which it is subject and to the direct contact with the crankshaft. This leads to faults due to metal-to-metal contact due to friction and sliding.

2. Experimental details

2.1 Visual examination

The tribological characterization of the connecting rod bearings was carried out once the worn bearings were extracted from the internal combustion engine. It was possible because the car engine was adjusted. It is important to mention that these bearings had a service life of approximately 8 years. In **Figure 2a–c**, it is possible to observe the complete crankshaft, the connecting rods, and a few worn bearings after disarming.

Firstly, small samples were extracted from worn bearings, approximately 1 cm^2, from different areas to perform measurements of hardness, roughness, chemical analysis of the materials that compound these mechanical components, and of course to identify the wear mechanisms. Micrographs of the damaged surfaces and identification of the different chemical elements were obtained using a scanning electron microscope (SEM) Quanta 3D FEG (FEI) equipped with an energy-dispersive X-ray spectrometer (EDS). In addition, roughness profiles of worn bearings after real service from different zones are obtained. **Figure 3** shows the damage zones of the connecting rod metals that were slightly affected by the contact with the rod bearing journals from the crankshaft. It is assumed that in this particular case, the lead and tin layer was not completely detached, and the bearing only suffered normal wear due to

(a)

(b)

(c)

Figure 2.
Adjust of car engine: (a) and (b) crankshaft and (c) connecting rods and bearings.

friction, applied load, contact speed, contact temperature, and fine abrasive wear by small strange particles that could be the same oil or inserted debris.

On the other hand, the wear produced in the bearings shown in **Figure 4** is very intense. In the bearing on the left, it is possible to observe that the lead and tin layer was completely torn off in the central part and a dark color was clearly seen, which was due to the oxidation wear that occurred in the steel layer that is the base layer of the bearing. In addition, the bearing on the right exhibited pronounced wear, although in this case the reddish-orange copper layer was clearly located. Here, it is possible to identify wear mechanisms such as sliding wear that became high abrasive damage (scoring) for high loads, fatigue wear due to constant contact with foreign particles, and high temperature.

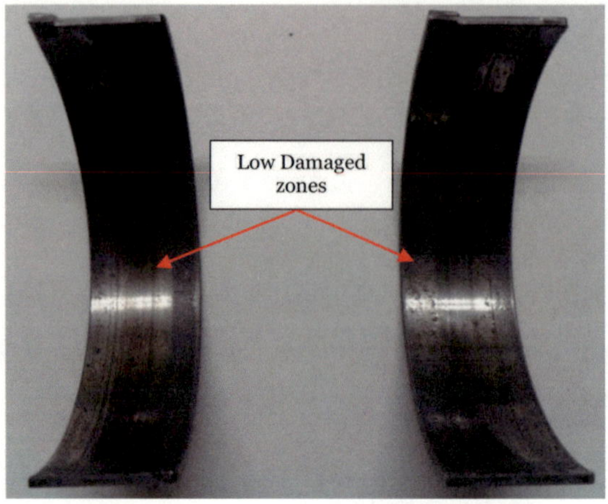

Figure 3.
Lower wear damage on bearings.

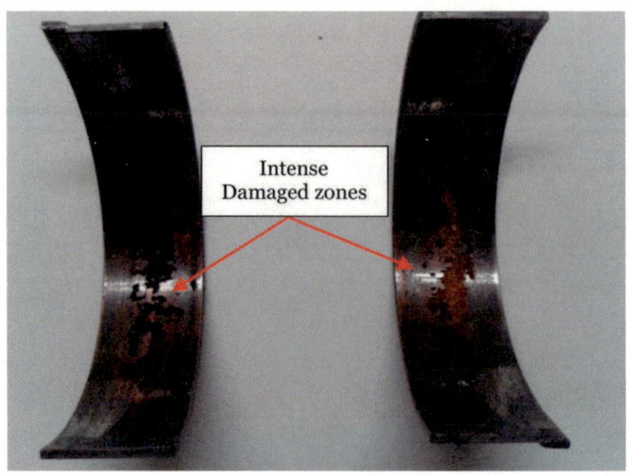

Figure 4.
Pronounced wear damage on bearings.

2.1.1 Hardness measurements

Hardness tests were carried out using a nanoindentation tester (TTX-NHT, CSM instruments). Ten hardness data from three different areas of the cross section of the bearing were calculated for each section, and an average value was obtained. The sections that were analyzed are shown in **Figure 5**. The applied load was 100 mN, the loading and unloading speed was 200 mN/min, and the duration of the measurement was 10 sec. In addition, the Young's modulus of the three different areas of the bearing was calculated as indicated in **Figure 5**. The microhardness results are presented in **Table 1** and **Figure 6**; the highest value corresponds to the outer layer of steel that is the one that makes direct contact with the rod body. In respect to the hardness of the central part, the values are very close to those of the outer layer, although this could be due to the use of copper or aluminum alloys which leads to a hardness increase. Of course, in the case of the inner layer, which is the one with the lowest hardness value, this corresponds to the Babbitt metal layer (lead-tin) that makes direct contact with the rod bearing journals from the crankshaft.

The results of the Young's modulus or modulus of elasticity (E) are shown in **Table 2** and in **Figure 7**. **Figure 8a–c** shows the traces of the pyramidal penetrator on the surface of the bearing in each of the examined areas.

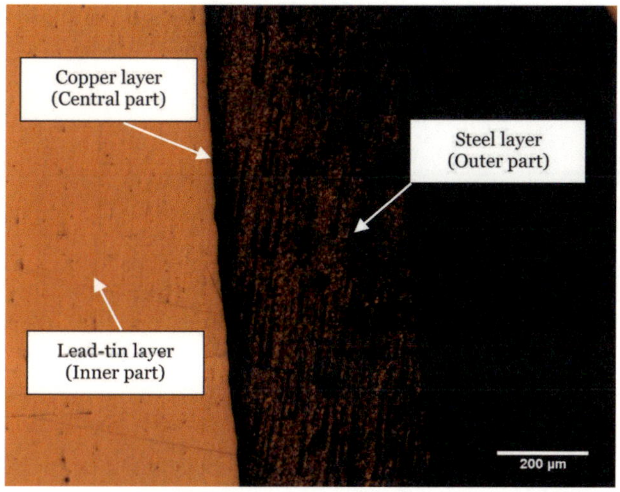

Figure 5.
Cross section of an unworn bearing.

Measurements (10)	Central part (HV)	Outer part (HV)	Inner part (HV)
Average value	192.97	196.61	21.89
Standard deviation	8.75	5.11	1.20

Table 1.
Hardness Vickers of the unworn bearing.

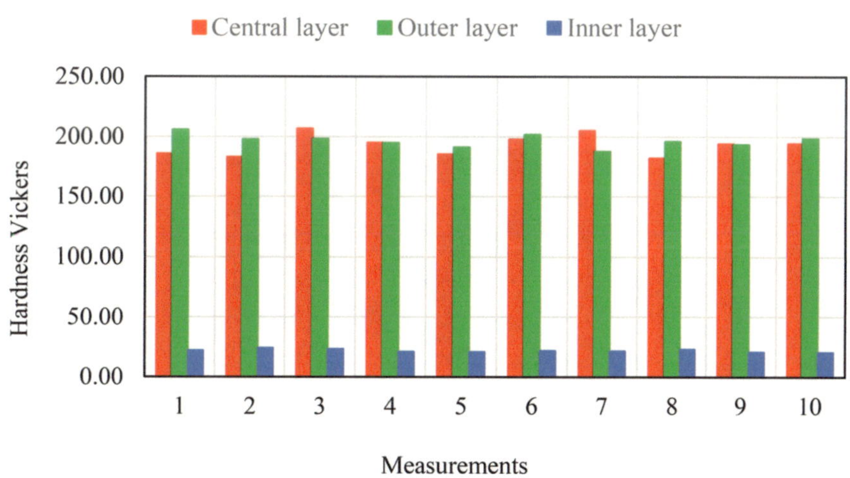

Figure 6.
Hardness measurements in different areas.

Measurements (10)	Central part (GPa)	Outer part (GPa)	Inner part (GPa)
Average value	209.56	211.19	37.67
Standard deviation	5.32	8.19	1.26

Table 2.
Young's modulus of the unworn bearing.

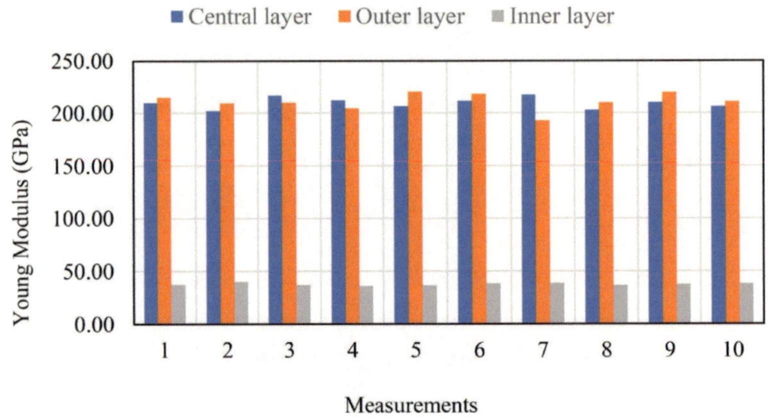

Figure 7.
Young's modulus measurements in different areas.

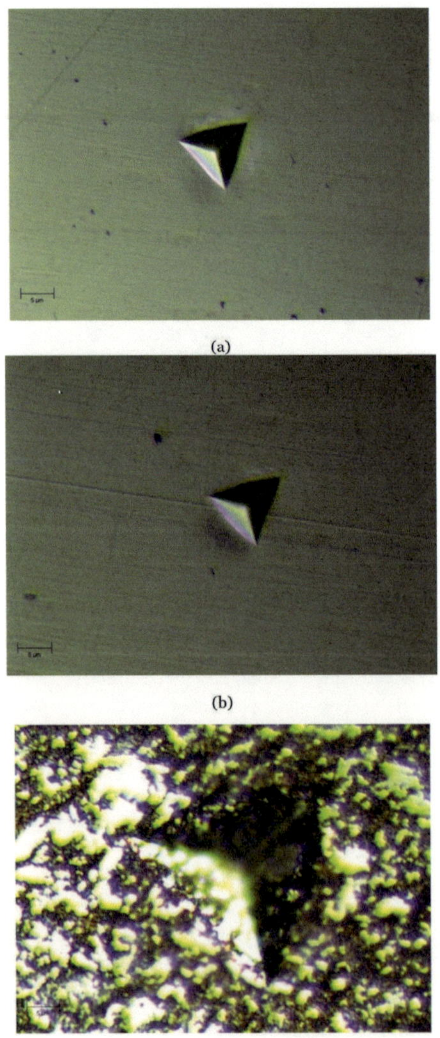

Figure 8.
Hardness Vickers indentations, (a) central part, (b) outer layer, (c) inner layer.

2.1.2 Roughness analysis

A roughness analysis was conducted in different wear zones of the connecting rod bearings; it was necessary to use a contour measuring instrument CV-500 (profilometer) located in the Laboratory of the Tribology Group at the National Polytechnic Institute. The profilometer is presented in **Figure 9**. Here, it is possible to observe the base where the piece of work is placed and the tip that slides on the surface; by means of this, it is possible to obtain the final roughness or depth results.

The photograph presented in **Figure 10a** shows how the surface of the damaged area changes in a consistent manner in the central part. In this particular case, the sliding wear was much less compared to **Figure 10b**, where the wear damage was more intense, with a high detachment of the outer layers, showing areas of greater depth due to the wear intensity.

2.1.3 EDS analysis

The chemical analysis was carried out using a scanning electron microscope model FE SEM JEOL JSM-7600F that can reach magnifications of 25× to 1,000,000× (see **Figure 11**). The unworn surface and the chemical analysis of the rod bearing layer that makes direct contact with the rod bearing journals from the crankshaft are shown in **Figure 12a** and **b**. Chemical elements such as C 20.86, Pb 60.05, Sn 3.89, and Sb 6.7 Wt% identified in the unworn surface.

On the other hand, once the rod bearings have been working during the actual service, the wear occurs on the surface as shown in **Figure 13a**. It was detected that the chemical composition was modified with respect to that presented with the bearing without any wear. Here, the Cu content was new, and this took place because the lead and tin layer was removed by the constant wear due to sliding, friction, and three-body abrasion that is generated by the wear debris insertion, contaminants, and foreign particles coming from of the oil filter or the atmosphere.

A comparison of the chemical composition of a rod bearing that suffered wear and an unworn bearing is presented in **Table 3**. In the chemical composition of the worn bearing, there is a reduction in the lead content (Pb) and an increase in the tin content (Sn). In addition, elements such as copper (Cu) and nickel (Ni) appeared after real service, which could be due to the foreign particles and contaminants produced during the contact with the crankshaft journal. This indicated that the layer of lead, tin, and antimony (metal Babbitt) was torn out from that area, so the copper content that is one of the intermediate layers in the bearing was clearly identified.

Figure 9.
Contour measuring instrument CV-500.

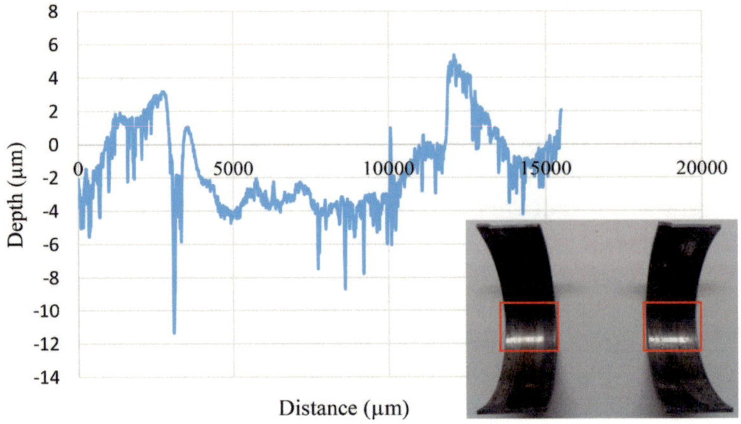

(a)

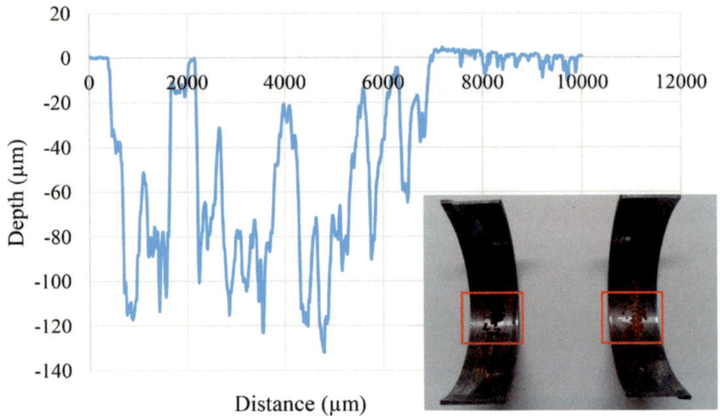

(b)

Figure 10.
Roughness profiles of the low wear area.

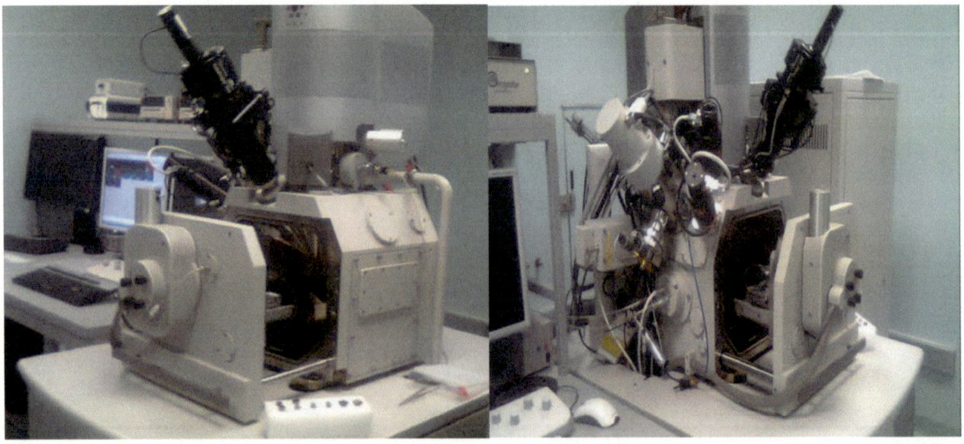

Figure 11.
Scanning electron microscope.

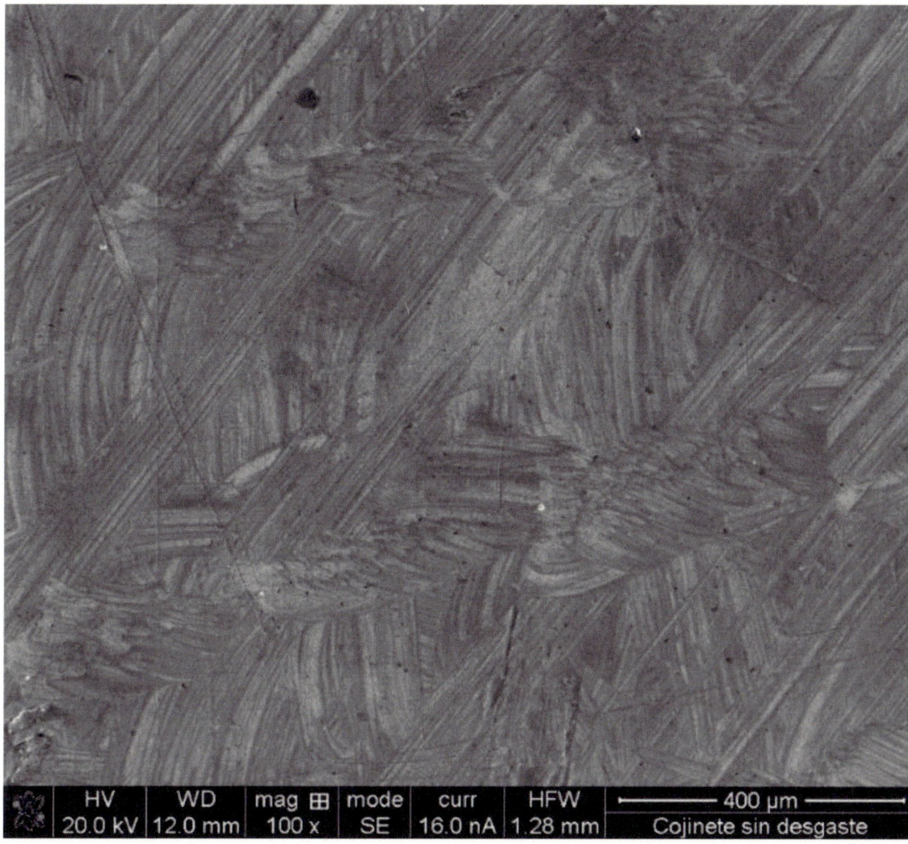

(a)

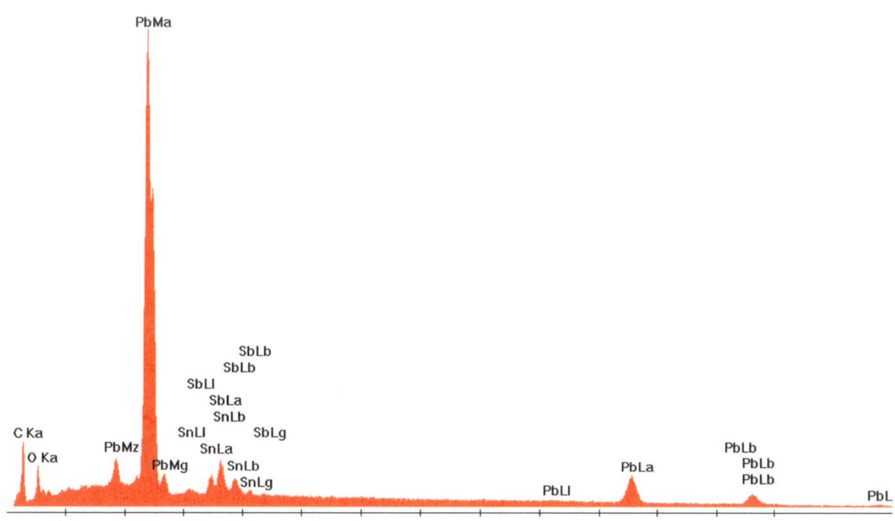

(b)

Figure 12.
(a) Unworn rod bearing surface and (b) EDS analysis.

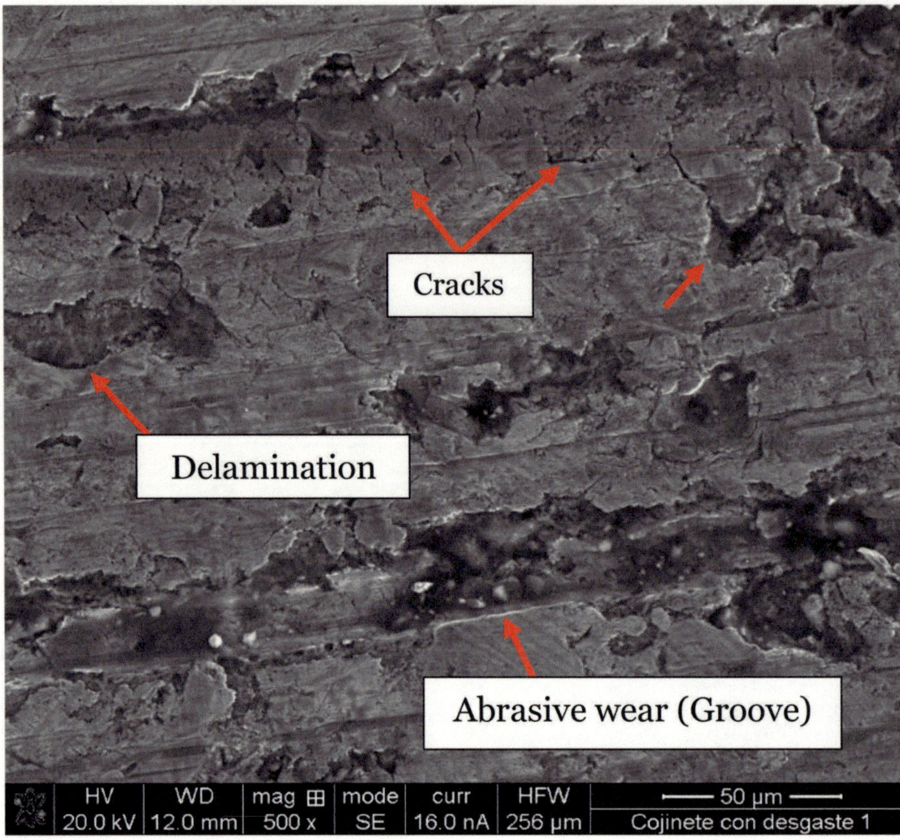

(a)

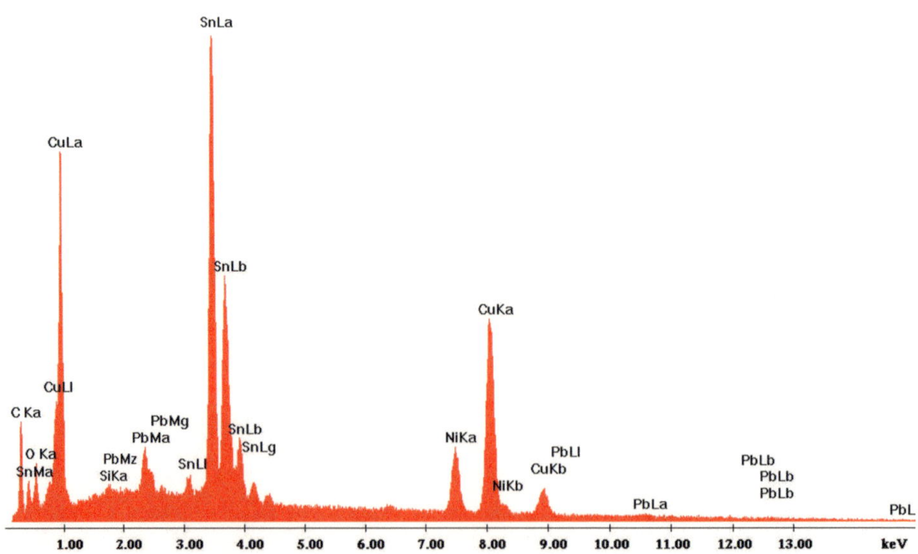

(b)

Figure 13.
(a) Worn rod bearing surface and (b) EDS analysis.

Element	Wt%	
	Worn bearing	Unworn bearing
C	25.33	20.86
O	13.5	8.5
Pb	6.8	60.05
Sn	40.72	3.89
Ni	3.21	6.7
Cu	10.43	—
Total	100	100

Table 3.
Comparison of chemical analysis of worn and unworn bearings.

3. Wear mechanisms and discussions

The wear mechanisms were identified by using a scanning electron microscope shown previously in **Figure 11** and an Olympus GX-51 optical microscope. **Figure 12a** showed the surface of the rod bearings before suffering any type of

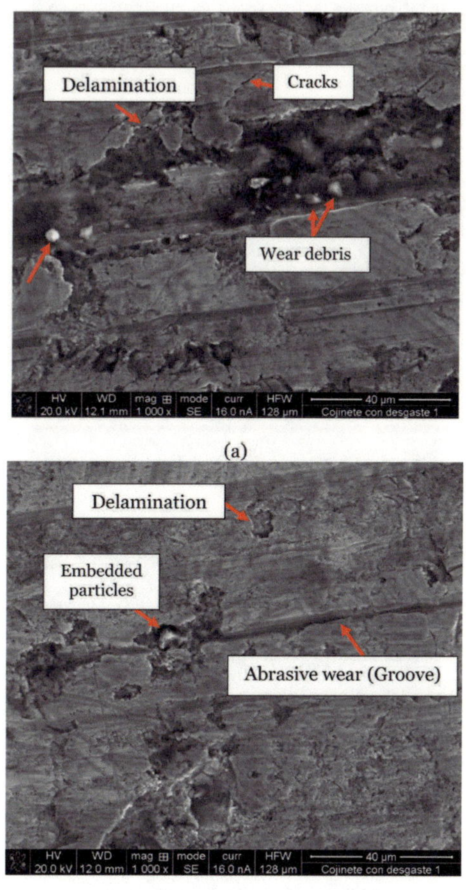

Figure 14.
(a) and (b) Worn rod bearing surface at 1000× in two areas.

damage. On the other hand, **Figure 14a** and **b** exhibited the damage caused in the bearings of the connecting rods after 8 years of actual service with proper and constant maintenance. On the bearing surfaces with the greatest wear (**Figure 4**), it was possible to observe cracks on the surface, delamination due to severe tearing of outer layers produced by sliding wear, and particles that were detached by the same action [2–5]. Some particles observed in the grooves are formed on the surface, and as the magnification of the damage area is increased, this wear mechanism is much clearer.

In the case of the bearing surfaces with the least wear (**Figure 3**), it is possible to mention that although the wear was less pronounced, delamination of the surface layers was observed, as the magnification of the damage zones increased (**Figure 15a–c**). In addition, it was possible to see particles embedded on the surface

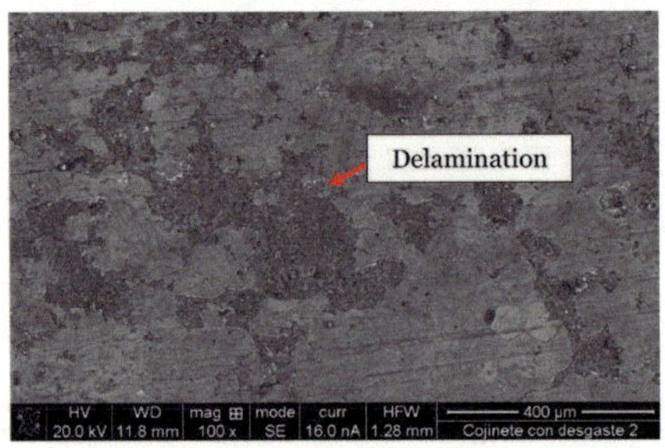

(a)

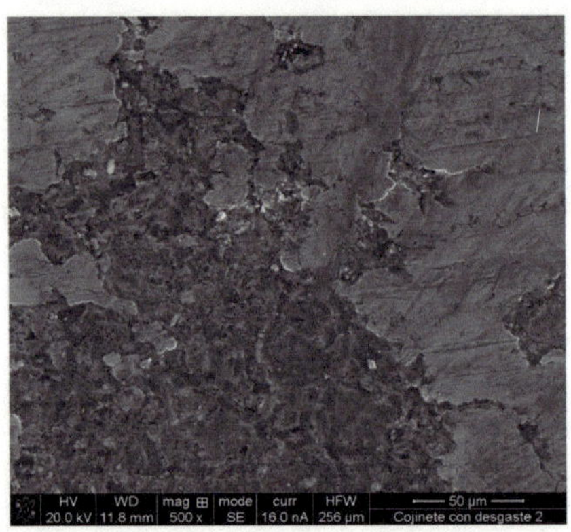

(b)

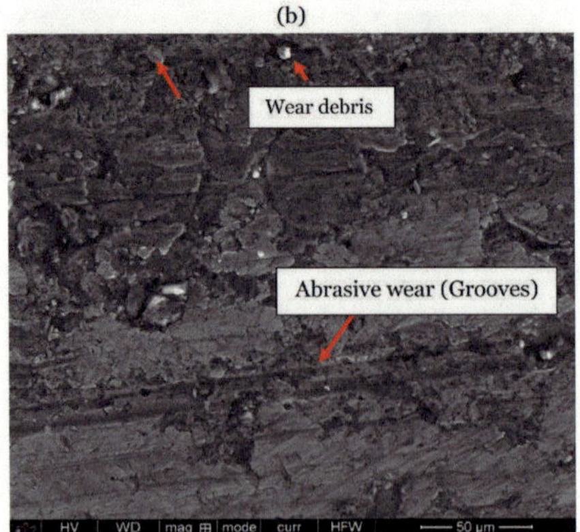

(c)

Figure 15.
(a)–(c) Worn rod bearing surface at 100× and 500× in different damaged areas.

Figure 16.
(a)–(d) Worn rod bearing surface in different damage areas.

by the same sliding action of the two surfaces in contact and grooves with less depth and intensity than those observed in **Figure 14a** and **b**.

In **Figure 16a–d**, it is possible to observe images of the wear damage in the bearings shown in **Figure 4**, caused by the intense tearing of a part of the first protective layer (Babbitt metal layer, composed of lead, tin, and antimony). An Olympus GX-51 optical microscope was used to obtain these micrographs. In a magnification of the same area, it was possible to observe that the fatigue damage was important since there are areas of severe material detachment, pitting action was also identified. In this particular, high abrasion wear (scoring) was the main wear mechanism to be considered by the action of foreign particles of various shapes and sizes on the surfaces that could lead to accelerate the wear process [5–7].

4. Conclusions

In this study, it was possible to understand more about the wear damage caused in the bearings of connecting rods. These mechanical elements are continually exposed to natural deterioration due to sliding contact with the crankshaft journals, by the incrustation of contaminating, foreign particles or even residues of the same wear that occurs; of course factors such as working temperature, lubricant degradation, bad assembly, and inadequate conditions can influence to cause serious wear problems in our cars.

In relation to the rod bearings that suffered higher wear damage (**Figure 4**), it was possible to observe that a Cu content appeared after the real service in the chemical analysis which confirmed the severe wear rate was caused by the absence of correct lubrication due to the long service life, roughly 8 years, and probable foreign particles (dust, wear debris) in the interface between the bearings and the shaft. The common sliding/abrasion action with particles between the two surfaces could be a major factor to inflict the wear intensity. In addition, the wear areas could lead to an uneven distribution of the loads, causing an abnormally high pressure area in some specific points on the bearing surface. In addition, Sn content was identified in the chemical results. Both protective layers were totally shifted due to the high abrasive action in some areas. SEM images showed a number of particles embedded on the bearing surface, large cracks, and clear abrasive wear due to the particles acting directly on the surfaces. Optical images presented wear areas with material totally detached from the surface.

On the other hand, the bearings with the lowest wear exhibited high polished areas on the surfaces in the sliding direction as shown in **Figure 3**, two factors could be significant, the absence of an oil film between the bearing and the shaft led to a metal-to-metal contact, and initial wear debris could act as polisher in the surfaces. In addition, pits on the surfaces were caused by a few wear debris inserted on the bearings. It can be inflicted by the contact pressure between the bearing and the shaft.

The roughness results showed the surface modifications due to the contact action between the rod bearings and connecting rod journal shafts. The main changes were identified in the high wear areas in the rod bearings in **Figure 4**, where a higher depth was obtained in the profiles. The roughness behavior in the rod bearings with the lowest damage was consistent in all the measured section. Finally, hardness tests were performed to show the differences between the inner part (lead-tin layer) and the two protective layers (steel and copper). The results presented a low hardness value of the inner part (21.89 ± 1.20) compared to the other two, 192.97 ± 8.7 HV, copper layer, and 196.61 ± 5.11 HV, steel layer, with close values.

Author details

Juan Rodrigo Laguna-Camacho[1*], Silvia M. Sánchez-Yáñez[2], Gabriel Juárez-Morales[1], Maria I. Cruz-Orduña[1], Luz M. Ramos-González[1], Cristóbal Cortez-Domínguez[1], Roberto L. Cabrera-Santiago[1] and Javier Calderón-Sánchez[3]

1 Faculty of Electrical and Mechanical Engineering, Universidad Veracruzana, Veracruz, Mexico

2 Facultad de Ciencias Biológicas y Agropecuarias, Universidad Veracruzana, Veracruz, Mexico

3 Universidad del Golfo de México Norte, Veracruz, Mexico

*Address all correspondence to: juanrodrigo.laguna@gmail.com

References

[1] Four Stroke Cycle Engines [Internet]. 2015. Available from: http://touchknow.blogspot.com/2015/09/four-stroke-cycle-engines.html [Accessed: 06 May 2019]

[2] Mustafa N, Cengiz O, Ibrahim C. Wear behavior of a crankshaft journal bearing manufactured by powder spraying. Materials and Design. 2008:2047-2051. DOI: 10.1016/j.matdes.2008.04.020

[3] ASTM standard, G76-95 (1995), standard practice for conducting erosion tests by solid particle impingement using gas jets. In: Annual Book of ASTM Standards; Vol. 03.02, Philadelphia, PA: ASTM; 1995. pp. 321-325

[4] Stachowiak GW, Batchelor AW, editors. Engineering Tribology. 3rd ed. USA, UK: Elsevier Butterworth-Heinemann; 2005. 482 p. Hardcover ISBN: 9780750678360

[5] Hutchings I, Shipway P, editors. Tribology: Friction and Wear of Engineering Materials. 1st ed. United States: Elsevier Butterworth-Heinemann; 1992. 165 p. Paperback ISBN: 9780340561843

[6] Engine Bearing: Failure Analysis and Correction [Internet]. 2019. Available from: https://www.mahle-aftermarket.com/files/local-media-north-america/product-files/ceb-2-1114-engine-bearing-failures-brochure.pdf [Accessed: 15 July 2019]

[7] Major Causes of Bearing Failure [Internet]. 2014. Available from: http://www.speedperf6rmanc3.com/content/Engine%20Bearings.pdf [Accessed: 15 July 2019]

Chapter 8

Technological Control on the Heredity of Operational Quality Parameters

Alexey G. Kolmakov, Mikhail L. Kheifetz, Nikolay L. Gretskiy and Gennady B. Prement

Abstract

Technological heredity of operational properties in the processes of manufacturing, operation, and restoration of machine parts is proposed to describe the graph reflecting the coefficients of transmission and mutual influence of physical, mechanical, and geometric parameters. The technological control methods of the heredity of operational properties of parts are considered, including the following: measuring parameters of the most critical parts; determining technological heredity mechanisms on the basis of transfer coefficients and mutual influence of operational properties; analyzing technological barriers during intensive effects of energy flows; and developing measures for controlling technological processes. The technological heredity of the operational quality parameters in the process of recovery and processing of the bearing journals and cams, as well as wearing of the camshaft working surfaces over admissible limits are considered. According to the findings, it is recommended: to control the deformation of the part after machining operations; to eliminate the editing operation after heat treatment; to use a combination of methods and a combination of technological effects in recovering the part surfaces with wear exceeding the maximum permissible values. The need for surfacing and subsequent tempering processes to ensure consistently high physicomechanical properties of coating materials and strictly regulate the modes of surface finishing is noted.

Keywords: technological heredity, operational properties, processed material, part surfaces, physicomechanical and geometric quality parameters, wear limit, engine camshaft, surfacing recovery, bearing journals, cams, mechanical treatment, internal combustion engine, crankshaft, camshaft

1. Introduction

The heredity in engineering technology is understood as the phenomenon of transferring the properties of the processed object from previous operations and transitions to the next ones, which further affects the operational properties of machine parts [1, 2]. The carriers of heredity information are the processed material and the part surfaces with all the variety of parameters describing them. Information carriers are actively involved in the technological process and in operation, going through various modes and transitions, experiencing the effects of technological factors [3, 4].

In the technological chain and at the stage of operation, there are some kinds of "barriers." Some technological factors cannot overcome them, and in this case, they do not affect the final properties of the object. Other factors pass through such "barriers," but their influence on the final properties significantly decreases [1, 3]. The most significant "barrier" is thermal operations, as well as operations accompanied by surface deformation and hardening, as they change the microstructure of the processed material and the microgeometry of the formed surface, leading to the deformation of the part and distortion of its shape. During these operations, various surface defects, such as structural heterogeneity, pores, and microcracks, can develop or heal. Consequently, it is possible to control the process of technological and operational heredity so that properties that positively affect the quality of the part are maintained throughout the entire technological process, while properties that affect negatively are eliminated at the beginning of the process [4, 5].

A distinctive feature of existing approaches to the definition and prediction of quality indicators for engineering products is the use of the superposition principle, according to which each of the existing technological factors is independent of the others, and the result of joint action is determined by their partial sum represented in one form or another [3, 4].

However, the technological systems are multiply connected, and the production objects are characterized by nonlinearity, irreversibility, and disequilibrium. The application of the superposition principle reduces the multiply connected interactions, which take place in the technological systems, to simply connected interactions, ignoring the mutual influence of technological factors [4, 5].

With the growth of requirements for the quality of machining parts' surfaces, methods for determining and predicting quality based on the principle of superposition become of little use, since the effect of the mutual influence of factors is comparable with the results of their direct impact. The process of ensuring the product properties should be considered as a set of interacting processes, changing and preserving the properties [6, 7].

In the repair industry, the transfer of physicomechanical and geometric quality parameters when performing various machining, welding, surfacing, heat treatment, surface hardening, and other operations determines the structure of the material and the surface layer of the part [2, 6, 7]. Therefore, the negative effects associated with the technological heredity should be taken into account when forming the technological process [1, 2, 7].

2. The synergetic concept of the state of a thermodynamic system

It is reasonable to determine the dominant processes of structure formation under intense effects in terms of synergetics using the concept of a mode in the distributions of continuous random variable of controlled parameter [8].

By a mode, we mean such a value of the parameter at which the density of its distribution is at a maximum. According to the synergetic concept, stable modes adjust to the dominant unstable modes and consequently can be excluded. This leads to a drastic reduction in the number of controlled parameters—degrees of freedom of a thermodynamic system. The remaining unstable modes can serve as the parameters of order that determines the processes of structure formation [9].

The equations of state of the thermodynamic systems, resulting from the reduction of the parameters, are grouped into several universal classes of the form [8, 9]:

$$\frac{\partial}{\partial \tau} U = G(U, \nabla U) + D \nabla^2 U + F(\tau), \quad (1)$$

where *U*—a controlled parameter; τ—current time; *G*—a nonlinear function of *U* and may be *U* gradient; *D*—the coefficient that describes either the diffusion when its value is real or the propagation of waves when its value is imaginary; and *F*—fluctuating forces due to the interaction with the environment and dissipation inside the system.

Equations of this type are similar to the equations describing phase transitions of the first and second kind. In accordance with the synergetic concept, phase transitions occur as a result of self-organization, a process which is described by three degrees of freedom, corresponding to the order parameter (O), its conjugate field (F), and control (C) parameter [9].

To use a single degree of freedom—the order parameter—is possible only for describing the quasi-static phase transformation. In the systems which are far from being in thermodynamic equilibrium, each of these degrees of freedom acquires an independent significance.

The processes of self-organization in them result from the competition of the positive feedback of the order parameter with the control parameter and the negative feedback with the conjugate field. Consequently, except for the process of relaxation, at the equilibrium state over time τ^p with two degrees of freedom, a self-oscillating mode can be realized, and with three degrees, a transition into a chaotic state can take place [8, 9].

Thus, the state of thermodynamic systems under intense treatment and operation is characterized by a number of modes [10, 11]:

1. *memorizing*—it is determined by a "frozen" disorder in the transition from a disordered state and is implemented when the time of the order parameter relaxation is much smaller than any other time ($\tau_O^p << \tau_C^p$ and $\tau_O^p << \tau_F^p$);

2. *relaxation*—it is realized when the time of relaxation of the order parameter is much greater than the relaxation time of the remaining degrees of freedom ($\tau_O^p >> \tau_C^p$ or $\tau_O^p >> \tau_F^p$);

3. *self-oscillation*—it requires the commensurability of the characteristic time of a change in the order parameter and the control parameter or the conjugate field ($\tau_O^p \gtrless \tau_C^p$ or $\tau_O^p \gtrless \tau_F^p$); and

4. *stochastic*—it is characterized by a strange attractor and is possible if all of the three degrees of freedom are commensurable ($\tau_C^p \gtrless \tau_O^p \gtrless \tau_F^p$).

The dominating processes of structure formation are determined by the intensity of energy and matter transfer in nonequilibrium thermodynamic systems. The stability of structure formation is provided by the control of the stability of the processes of intensive processing and operation through the use of positive and negative feedbacks [10, 11].

2.1 Thermal treatment of metals and surfaces

The purpose of any thermal treatment processes is to provide a desired material structure by heating (or cooling) it up to a certain temperature and subsequently changing it [12]. The mode of thermal treatment is typically characterized by the following basic parameters: the temperature of heating and holding time and the speed of heating and cooling of the material [13].

All the types of thermal treatment, according to Bochvar [14], are divided into four main groups of operations, which in terms of the synergetic concept of structure formation can be associated with the modes of the thermodynamic system behavior.

The modes are defined by relaxation times τ^p, which refer to (i) the order parameter at cooling, (ii) the structure formation parameter conjugated with the previous one, and (iii) the control parameter for thermal treatment, namely heating. The presence of two degrees of freedom determines a thermal cycle, while three degrees of freedom denote a cycle with phase transitions.

As a result, groups of operations of thermal treatment are implemented [13, 14]:

1. *hardening*—heating above the transformation temperature followed by apid cooling to obtain a structurally unstable state;

2. *tempering*—heating of the hardened material below the transformation temperature followed by cooling to obtain a more stable structural state;

3. *annealing of the first kind*—heating the material in an unstable state after the pretreatment followed by slow cooling, resulting in a more stable structural state; and

4. *annealing of the second kind*—heating above the transformation temperature followed by slow cooling to obtain a structurally stable state.

2.2 Plastic deformation and metal forming

Phase transformations used in thermal treatment are primarily caused by a change in temperature, but varying the other thermodynamic factor—the pressure—it is possible to obtain structural changes that do not occur at constant pressure [12, 13].

According to the synergetic concept of structure formation, types of materials forming, like the types of thermal treatment, can be divided into four main groups of operations related to the modes of behavior of a thermodynamic system [8, 11].

Forming modes are also determined by time τ^p: of the order parameter during relaxation (stress relieving), structure formation parameter conjugate with it, and the parameter of mechanical processing control—pressure. Two degrees of freedom determine cyclic hardening, and three—stochastic cold hardening and destruction.

Hence, the following metal forming processes corresponding to different sections of the generalized curve "strain-stress" are [8, 11]:

1. *impact*—local or uniform pressure to form a state of stress and deformation structures or destruction;

2. *stress relaxation*—no pressure after preloading accompanied by the removal of internal stresses and formation of more equilibrium structures;

3. *cyclic cold hardening*—the creation of hardening deformation structures by cyclic formation of the stress state as a result of the application and removal of the load; and

4. *stochastic cold hardening*—the creation of hardening deformation structures by aperiodic formation of stress state as a result of stochastic loading.

3. Quality indicators of machine parts surfaces

The quality indicators of engineering products, which are the main ones, are divided into two categories [15, 16]: the first category includes those that are characterized by heredity phenomena related to the properties of product materials; and the second category comprises the quality indicators related to the geometrical parameters of their surfaces.

Indicators of both categories in multiply connected technological and operational environments mutually influence each other. Geometrical product parameters, such as product configurations and sizes, can influence the stresses distributed in the base material and surface layers. On the contrary, the stresses generated during the technological hardening process and operation stages may, over time, lead to changes in the geometrical parameters of the parts. This testifies to the interconnection and conditionality of the phenomena accompanying the technological and operational processes.

The most complete heredity of the main quality indicators is revealed when considering the sequence of processes from the synergistic positions of the joint action of technological factors with the mutual influence of indicators [10, 11].

The initial quality indicators for machine parts at various scale levels (**Figure 1**) vary during operation [7, 10]. The exceptions are the residual stresses and the structure of the base material, which can be maintained until the rubbing surfaces of the parts are completely destroyed. In most cases, already during the period of running-in, the roughness and structure of the surface relief significantly change. The waviness and structure of the surface layers of the part change with steady wear, and the geometric shape of the friction surface remains within the allowable values adopted during manufacture almost to the end of the friction unit service, if its performance is assessed by accuracy parameters [2, 5, 7].

Reducing the sensitivity of technological and operational environments to the changes in the conditions for the implementation of production modes and the use of products allows to carry out the directional formation of quality indicators in the life cycle of engineering products for the least cost [2, 16]. The functional models of multiply connected technological environments allow, depending on the

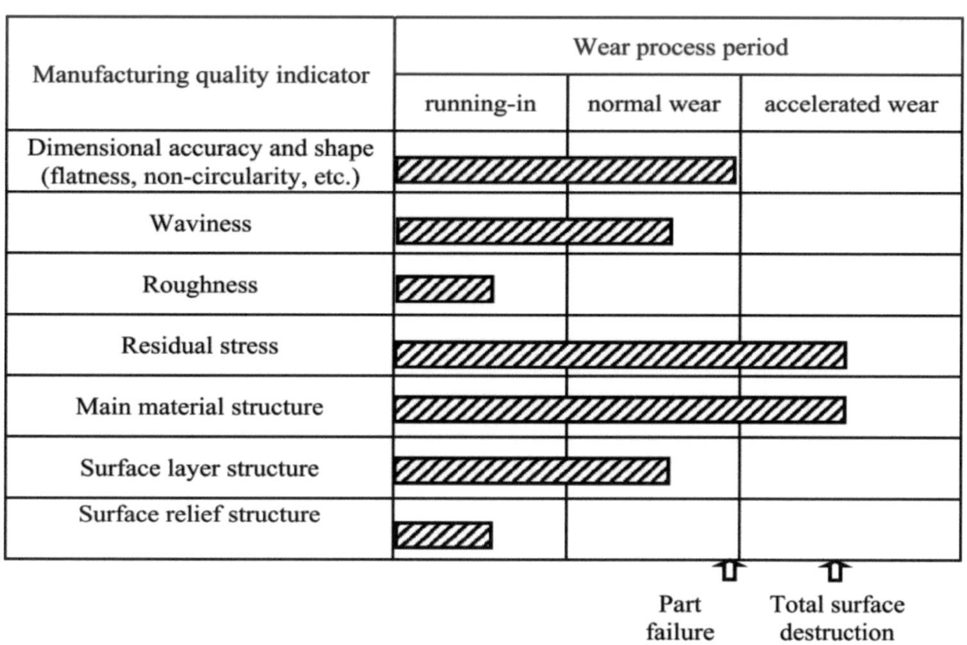

Figure 1.
Diagram of changes in initial quality indicators of product surface during operation (shaded areas characterize the duration of preservation of the initial values in the geometric parameters, residual stresses, and the material structure within permissible deviations).

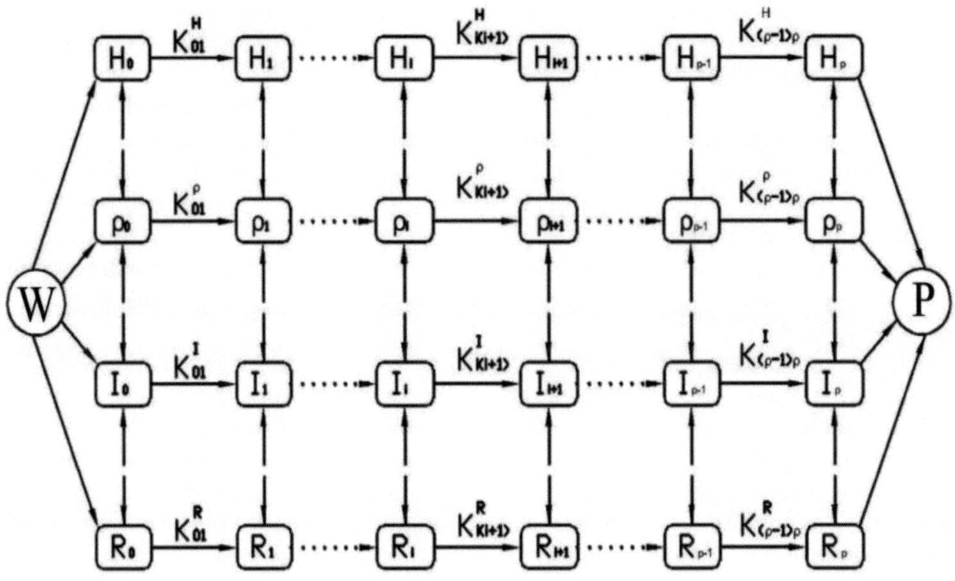

Figure 2.
A detailed graph of technological and operational heredity with a set of quality indicators.

formulation of the problem being solved, to reduce its dimension by highlighting a number of essential connections and suppressing insignificant connections while maintaining correctness and adequacy [6, 11].

3.1 The structure of the relationship of inherited properties

The technological process of manufacturing and operating parts can be represented as a graph, highlighting the procurement, draft operations, finishing operations, as well as performance stages [7].

The graph, as a rule, is oriented, and the quality parameters are interconnected (**Figure 2**).

The initial vertex of the graph, when describing the technological process, is a workpiece W, the final vertex is the finished part P in operation. Oriented edges of the graph show the transfer of operational properties of the part during processing.

The edge transfer is described by the heredity coefficient K, reflecting a quantitative change in the property and equal to the ratio of the previous S_j and subsequent S_{j+1} property values [4]:

$$K = S_j / S_{j+1} \tag{2}$$

In addition to the direct transfer of properties (**Figure 2**) with technological and operational heredity, it is advisable to evaluate their interaction (**Figure 3**).

3.2 Main inherited quality indicators

To identify the main quality indicators inherited in operation, through the control of which it is advisable to manage the technological process, the ABC analysis was performed (**Figure 4**), highlighting the reasons for the change in the initial geometric parameters of the surface and the physicomechanical characteristics of the material during operation [5].

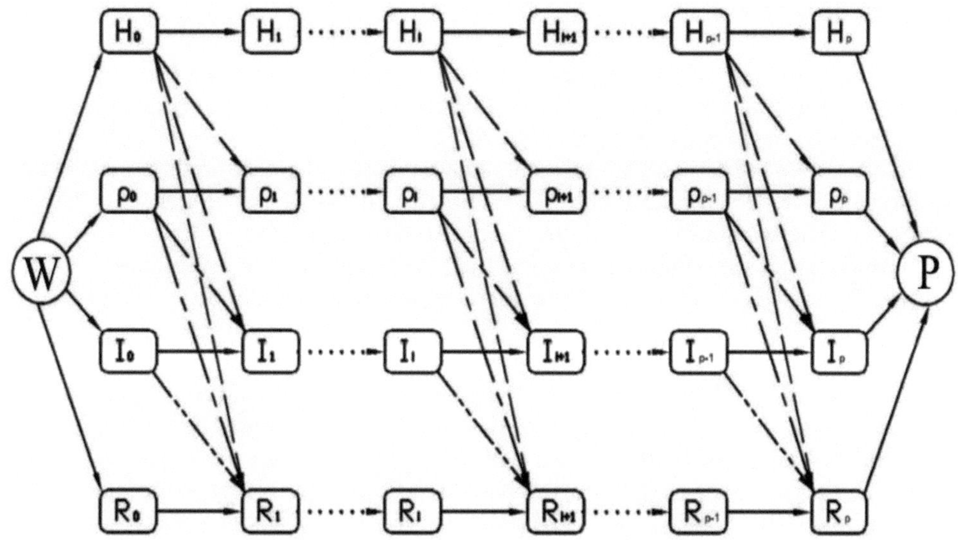

Figure 3.
The graph of technological and operational heredity reflecting the mutual influence of physicomechanical and geometric quality indicators.

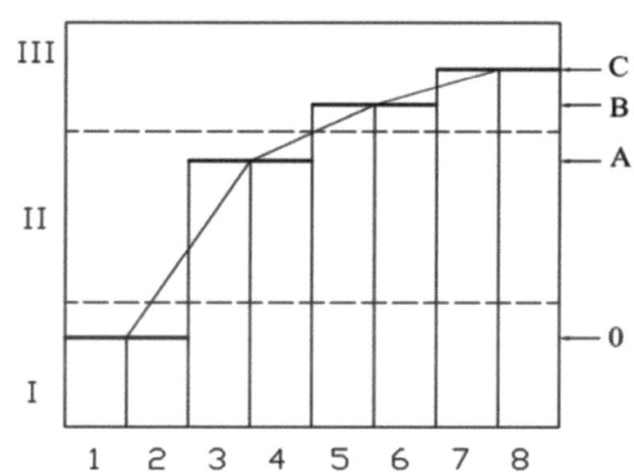

Figure 4.
ABC analysis of changes in operation process (I–II) of initial quality indicators (1–8): 0—the surface formation; A—the change in contact loads; B—part failure; C—complete destruction of the surface; I—running-in; II—normal wear; III—catastrophic wear; 1—the surface roughness; 2—the surface relief structure; 3—the surface waviness; 4—the structure of the surface layers; 5—the surface shape; 6—the dimensional accuracy; 7—residual stresses; and 8—the structure of the main material.

The ABC analysis showed that in most cases, already during the running-in period (I), the roughness (1) and the surface relief structure (2) change significantly. The waviness (3) and the structure of the surface layers (4) change with steady wear (II). The accuracy of dimensions (5) and the geometric shape of the surface (6) remain within acceptable limits even at the beginning of the catastrophic wear stage (III). Only residual stresses (7) and the structure of the main material (8) can be maintained until the rubbing surfaces are completely destroyed [7].

Therefore, to study the heredity, we selected the operatively and least laboriously controlled physicomechanical geometric quality indicators from the initial and final groups (0–C). At the same time, special attention was paid to indicators (5, 6) undergoing significant changes at the beginning of catastrophic wear (B) and related both to the physicomechanical characteristics of the material (7, 8) and to the geometric parameters of the surface relief (1, 3).

3.3 Methods of research

The study and management of the technological and operational heredity by the proposed method of quality parameters control was carried out for the parts responsible for the product life [10, 17].

Measurements of hardness HRC, shape deviations ρ, dimensional accuracy *IT*, and surface relief *Ra* were carried out on a batch of parts. It was divided into 10 groups, and the arithmetic average of the group was taken as the calculated value.

On the basis of the calculated results, the heredity transfer coefficients K^{H}, K^{ρ}, K^{I}, and K^{R} were determined for the graph in **Figure 6**, and the coefficients of the technological effect heredity $K^{H\rho}$, K^{HI}, K^{HR}, $K^{\rho I}$, $K^{\rho R}$, and K^{IR} were done for the graph in **Figure 3**.

To assess the heredity of the technological route, the resulting coefficients K_r were calculated, equal to the product of the corresponding coefficients for the operational quality parameters throughout the entire sequence of operations. To determine the degree of heredity influence on various technological operations, the comparison coefficients K_c, equal to the ratio of mutual influence coefficients on the previous and subsequent operations, were calculated [18].

4. The heredity quality parameters by recovering

The technological process of recovery and hardening of the camshaft can be divided into the stages of flaw detection; shape recovery and hardening; final surface treatment; running on the stand; and further operation. In this regard, the study of all quality parameters was carried out not after each individual operation, but after the selected stages.

The parameters were measured on the surfaces of the bearing journals and cams since these surfaces are friction surfaces and are constantly in contact with other surfaces. On the surface of the bearing journals, friction forces are constantly acting, and the cams undergo cyclic loading [2, 7].

The measurement results were entered into the tables in which the data on the number of each journal or cam are vertically arranged, and the data on the classes are presented horizontally. The data were designed in such a way that it was possible to analyze not only the measurement results of different classes of parts, but also the differences in the data for different numbers of journals and cams throughout the entire length of the shaft. On the basis of the tabular data, the graphs were built (**Figures 5–11**) showing the changes in the quality parameters for selected stages of the technological process of recovering the UMZ-4173 engine camshaft. The graphs analyzed the processes of changing the operational quality parameters and found their patterns.

According to the experimental data, the transfer coefficients of the technological heredity, equal to the ratio of individual quality parameters before and after the operation, and the mutual influence coefficients of various quality parameters during processing and operation were calculated. The calculations were carried out for bearing journals (no. 1–5) and cams (no. 1–16) in classes (no. 1–10) allocated (stratified) depending on the wear degree of the working surfaces of the UMZ-4173 engine camshafts (**Tables 1–4**).

In experimental studies of the operational parameters of the UMZ-4173 engine camshaft, the hardness of the material was taken as the main physicomechanical parameter. The radial runout, dimensional accuracy, and surface roughness were taken as the main geometrical parameters.

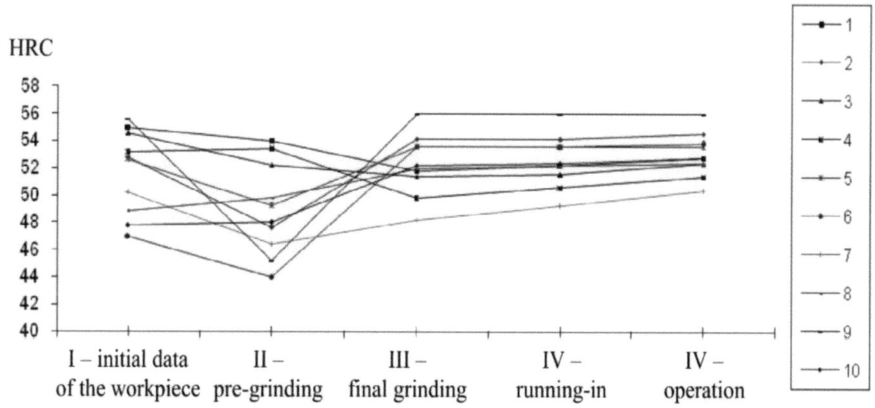

Figure 5.
The dependences of the changes in hardness of bearing journals by classes (no. 1–10) of camshafts in operations and stages (no. I–V).

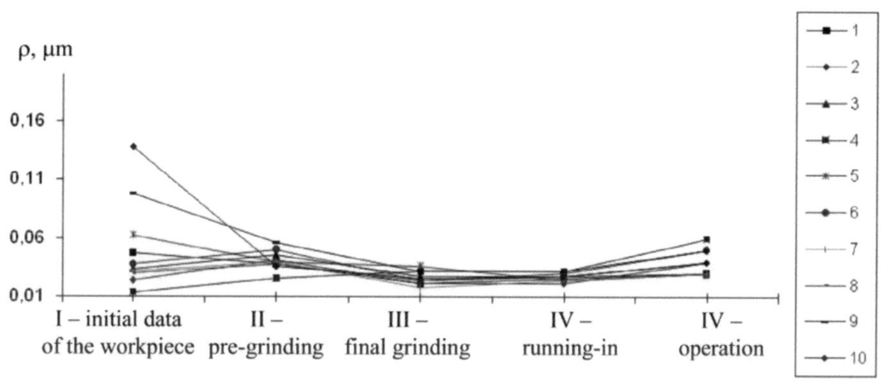

Figure 6.
The dependences of the changes in the radial runout of the bearing journals by classes (no. 1–10) of camshafts in operations and stages (no. I–V).

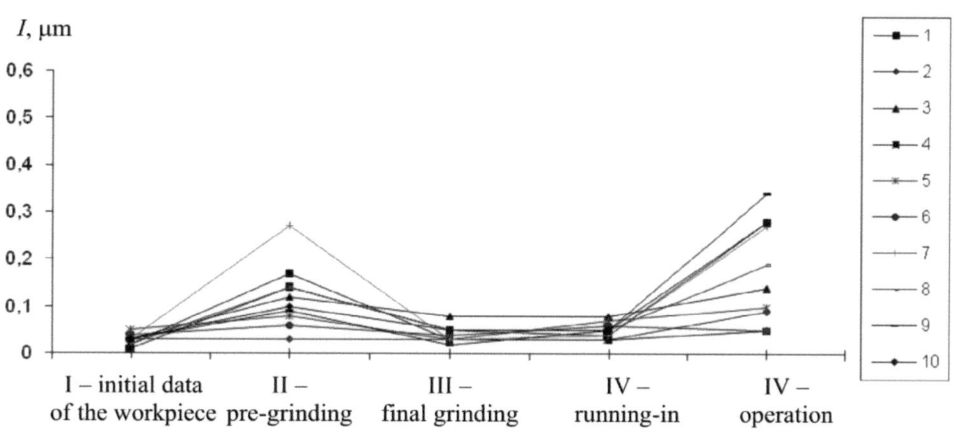

Figure 7.
The dependences of the changes in the dimensional accuracy of the bearing journals by classes (no. 1-10) of camshafts in operations and stages (no. I–V).

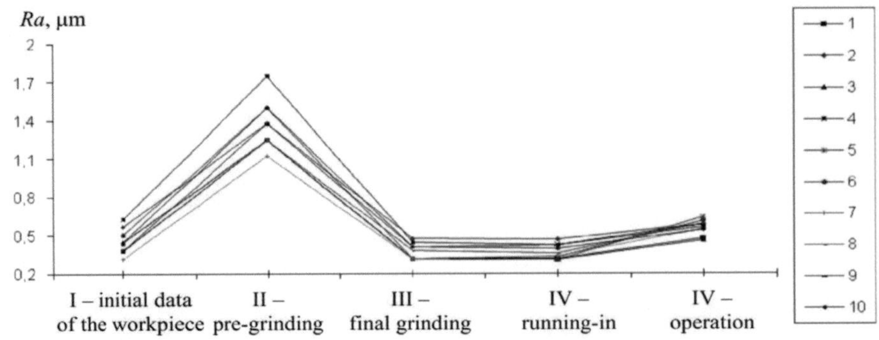

Figure 8.
The dependences of the changes in the surface roughness of the bearing journals by classes (no. 1–10) of camshafts in operations and stages (no. I–V).

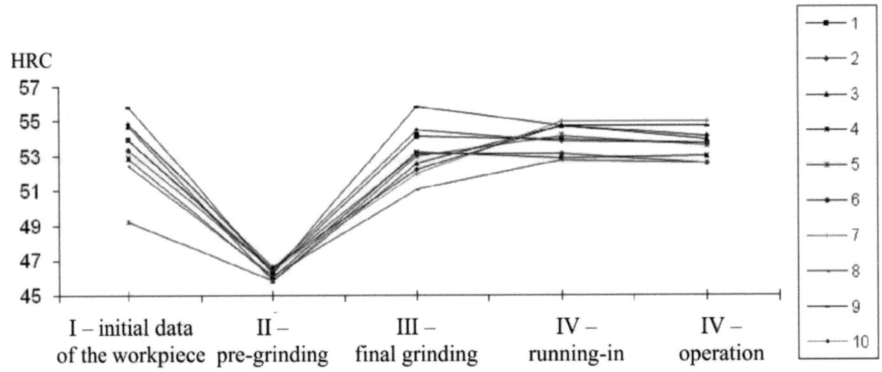

Figure 9.
The dependences of the cam hardness by classes (no. 1–10) of camshafts in operations and stages (no. I–V).

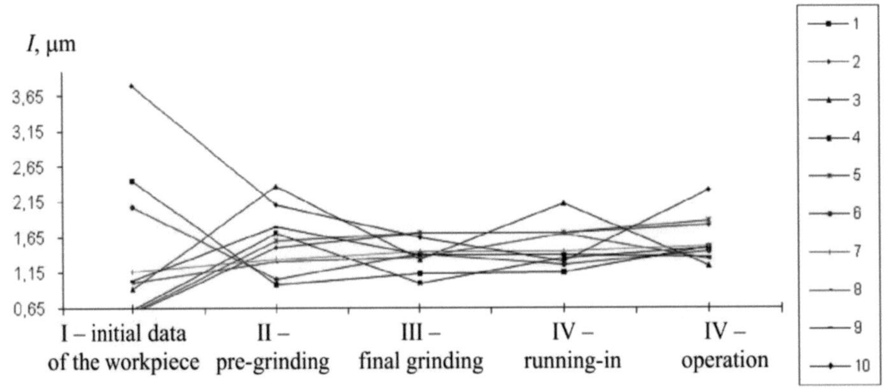

Figure 10.
The dependences of the changes in the dimensional accuracy of the cams by classes (no. 1–10) of camshafts in operations and stages (no. I–V).

The geometrical and physicomechanical parameters are interrelated; therefore, it is necessary to analyze the patterns of their changes in a complex in order to explain the mechanisms of the technological heredity associated with the interaction of properties. So, the radial runout of the surface during operation is greatly influenced by both the dimensional accuracy and its roughness, and with a

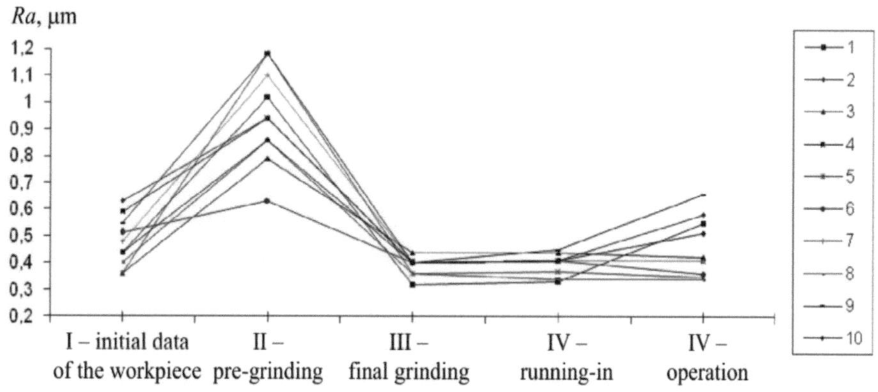

Figure 11.
The dependences of the changes in the surface roughness of the cams by classes (no. 1–10) of camshafts in operations and stages (no. I–V).

Recovery and operation	The quality parameters transfer coefficients			
	K^H	K^ρ	K^I	K^R
Workpiece—pre-grinding (K_1)	1.0568	1.2573	0.2500	0.3303
Pre-grinding—final grinding (K_2)	0.9369	1.5606	3.9024	3.6634
$K_{rg} = K_1K_2$	0.9901	1.9621	0.9756	1.2102
Final grinding—running-in (K_3)	0.9947	0.9635	0.9111	1.0092
Running-in—operation (K_4)	0.9913	0.6850	0.1800	0.6563
$K_{ro} = K_3K_4$	0.9860	0.6600	0.1640	0.6624
$K_r = K_{rg} \cdot K_{ro}$	0.9762	1.2950	0.1600	0.8016

Table 1.
The transfer coefficients K and the resulting heredity factors Kr for hardness H, shape deviations ρ, dimensional accuracy I, and surface roughness R of the camshaft bearing journals.

Recovery and operation	The coefficients of the mutual influence of quality parameters					
	$K^{H\rho}$	K^{HI}	K^{HR}	$K^{\rho I}$	$K^{\rho R}$	K^{IR}
Workpiece—pre-grinding (K_1)	1256.3107	323.5000	37.9806	0.3237	0.0380	0.0294
$K_{12} = K_1/K_2$	0.6771	0.2708	0.2885	0.3222	0.3432	0.0682
Pre-grinding—final grinding (K_2)	1855.3030	1194.6341	131.6667	1.0049	0.1108	0.4301
$K_{23} = K_2/K_3$	0.9724	1.0283	0.9283	1.7129	1.5463	3.8668
Final grinding—running-in (K_3)	1908.0292	1161.7778	141.8340	0.5867	0.0716	0.1112
$K_{34} = K_3/K_4$	1.4521	5.5260	1.5155	5.3528	1.4680	1.3882
Running-in—operation (K_4)	1314.0000	210.2400	93.5897	0.1096	0.0488	0.0801

Table 2.
The mutual influence coefficients K and the comparison coefficients Kc for hardness H, shape deviations ρ, dimensional accuracy I, and surface roughness R of the camshaft bearing journals.

Recovery and operation	The quality parameters transfer coefficients		
	K^H	K^I	K^R
Workpiece—pre-grinding (K_1)	1.1594	0.9061	0.5005
Pre-grinding—final grinding (K_2)	0.8698	1.1138	2.4593
$K_{rg} = K_1K_2$	1.0085	1.0092	1.2310
Final grinding—running-in (K_3)	0.9850	0.9365	0.9720
Running-in—operation (K_4)	1.0051	0.9503	0.8630
$K_{ro} = K_3K_4$	0.9900	0.8899	0.8389
$K_r = K_{rg} \cdot K_{ro}$	0.9984	0.8981	1.0326

Table 3.
The transfer coefficients K and the resulting heredity coefficients Kr for hardness H, dimensional accuracy I, and surface roughness R of the camshaft cams.

Recovery and operation	The coefficients of the mutual influence of quality parameters		
	K^{HI}	K^{HR}	K^{IR}
Workpiece—pre-grinding (K_1)	34.0339	56.5200	1.4453
$K_{12} = K_1/K_2$	1.0410	0.4714	0.3511
Pre-grinding—final grinding (K_2)	32.6943	119.8899	4.1169
$K_{23} = K_2/K_3$	0.9288	0.8949	1.1612
Final grinding—running-in (K_3)	35.2002	133.9736	3.5453
$K_{34} = K_3/K_4$	1.0365	1.1413	1.0976
Running-in—operation (K_4)	33.9623	117.3913	3.2299

Table 4.
The mutual influence coefficients K and the comparison coefficients Kc for hardness H, dimensional accuracy I, and surface roughness R of the cams.

relatively large surface runout, it is impossible to speak of high dimensional accuracy. The physicomechanical properties of the material have a significant impact on the geometrical parameters of the part.

According to the experimental data, the heredity mechanisms of operational quality parameters in the process of recovering the working surfaces of the UMZ-4173 engine camshafts were analyzed.

The hardness of the journal and cam surfaces at the stage of flaw detection gives a large scatter of data on the camshafts. This is due to the different condition of the shafts that came to overhaul. With an average hardness value of 55 HRC, the hardness of the shafts of no. 9 and 10 classes is significantly lower (44 HRC).

After restoration of a worn part by surfacing using a welding wire, the hardness drops sharply since the deposited layer without quenching does not have the hardness that the part had before the surfacing. But after heat treatment and subsequent finishing operations, the hardness is not only recovered, but in some cases reaches higher values than the initial workpiece had. This indicates that the hardening is carried out in full accordance with the technological process.

The classes of shafts, which at the initial stage had low hardness (no. 9 and 10), also have the lowest hardness values after the finishing treatment as compared to other classes. This obviously manifests the technological heredity.

Analyzing the geometric quality parameters, their dependence on the physicomechanical characteristics of the surface layer material should be taken into account. The initial radial runout of the camshaft bearing journals surfaces have a very large variation, which indicates a large wear of the bearing journals surfaces. After recovery operations, the runout values slightly stabilize (0.03–0.07 mm), but are still far from the required ones. After the final grinding, the values do not improve and more often remain at the same level. Moreover, in some classes, no. 10, the radial runout value deteriorates as compared to the initial one (from 0.01 mm on the initial workpiece to 0.03 mm on the recovered part).

These changes may be the result of editing operations that are carried out in the process of recovering parts. Editing a part, in addition to positive effects (elimination of residual bending, warping or twisting deformations), can also have negative effects on quality parameters.

The changes in the process of recovering the radial runout values in those classes of camshafts in which low surface hardness was noted (classes no. 9 and 10) is very ambiguous. This is a manifestation of the mutual influence of quality parameters in the technological heredity process.

When analyzing the dimensional accuracy of the camshafts surfaces, it should be noted that the changes in the accuracy of bearing journals and cams surfaces differ significantly. The bearing journals surfaces arrive for overhauls sufficiently rolled-in; therefore, a large deviation in the dimensional accuracy is not observed. The cams, on the contrary, have a large variation in the values of both the maximum cam size and the minimum one (i.e., the diameter of the cam base). This is due to varying degrees of wear on the cam surfaces.

After surfacing and turning operations, the scatter of the size values for bearing journals and cams is preserved and sometimes increases. But if, after finishing operations, the accuracy of the bearing journals is noticeably improved, although it does not reach the initial values, then the accuracy of the cams does not change, and sometimes (as in the case of class no. 10), it deteriorates compared to previous operations, which are also associated with the heredity phenomena.

The changes in the roughness of bearing journals and cams surfaces are unconventional for machining: after roughing operations, the roughness deteriorates in comparison with the initial condition, and in terms of finishing, it improves in comparison with the initial values. Moreover, the heredity processes are similar both on the bearing journals and cams, and the roughness parameters are less affected by other geometric or physicomechanical quality indicators.

5. Technological control of operational quality parameters

The study of experimental data allowed to determine the main dependences of the technological heredity of the physicomechanical and geometric quality parameters of the UMZ-4173 engine camshaft during the repair process. The transfer coefficients (**Tables 1** and **3**) show that the technological recovery and hardening processes are fundamentally different from the rational technological machining process. During machining in the manufacturing process, the harmful influence of the technological heredity is eliminated in the initial operations, while in the final operations, the heredity coefficients are stabilized.

When recovering during the repair process, both geometrical and physicomechanical parameters first deteriorate, and then they improve. However, in general, throughout the entire process, the physicomechanical characteristics are recovered, and the geometric, especially those associated with the surface microrelief, are even improved.

The mutual influence coefficients (**Tables 2** and **4**) allow to estimate the significance of both technological operations and technological factors in individual operations. Thus, the hardness of the material significantly affects the geometric parameters. For shape deviations, this effect is especially important in the initial operations. In other cases, it is stable in all technological transitions.

The geometrical parameters of the cylindrical surfaces of the bearing journals are weakly inherited; this is especially noticeable in the initial operations. Moreover, for the surface microrelief (its roughness), the recovery operations are technological "barriers" (since K^{pR} and $K^{TR} \to 0$). The further influence of the previous geometrical parameters on the subsequent ones is also not great and affects only the accuracy of processing.

The changes in the hardness of the bearing journals and cams of the UMP-4173 engine camshafts (**Figures 5** and **9**) show that the surfacing operations are technological barriers to the recovery of the working surfaces, and the final geometrical parameters of the surface quality are formed during finishing processing.

The high quality of repaired machines can be ensured by introducing new and traditional methods of recovery, hardening, and processing of machine parts [1, 2]. However, they have their own rational areas of application and do not always solve complex tasks of increasing the durability of products in specific operating conditions [6, 7].

Thus, the economical recovery of the extremely worn-out part surface to a given size is often not ensured with high quality parameters of hardening. Therefore, it seems rational to combine various methods of hardening and machining in the technological process of recovery, as well as various technological effects within the framework of the methods themselves [19–21].

In this regard, the technologies and equipment for combined hardening dimensional surface treatment of parts by applying ferromagnetic powders with electromagnetic welding in combination with surface plastic deformation (**Figure 12**) and welding of low-alloyed carbon wires in combination with rotary cutting have been proposed in order to recover the camshafts with varying wear degree of the working surfaces (**Figure 13**).

The combination of hardening, recovery, and surface treatment of parts in one technological process makes it possible not only to provide the necessary geometric characteristics of the surface during the recovery process but also to improve the physicomechanical properties of the surface layer material during hardening [19, 22, 23].

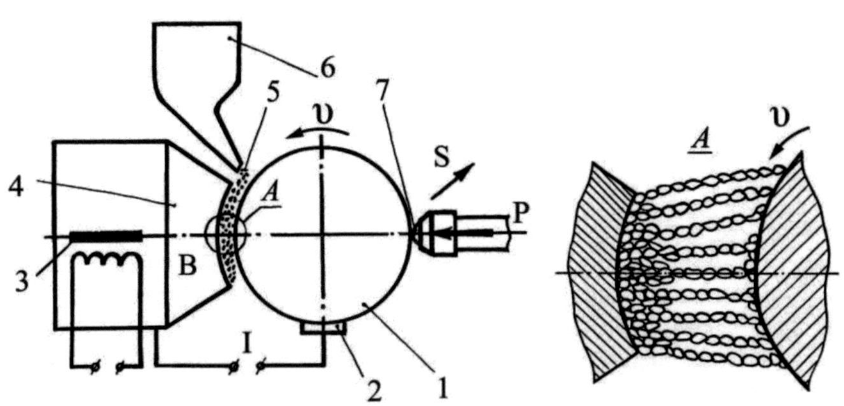

Figure 12.
Electromagnetic surfacing with surface plastic deformation: 1—workpiece; 2—sliding contact; 3—electromagnet; 4—pole piece; 5—ferromagnetic powder; 6—dosing device; 7–ball runner; υ—main speed; S—feed rate; P—deformation force; B—magnetic induction; I—electric arc current strength.

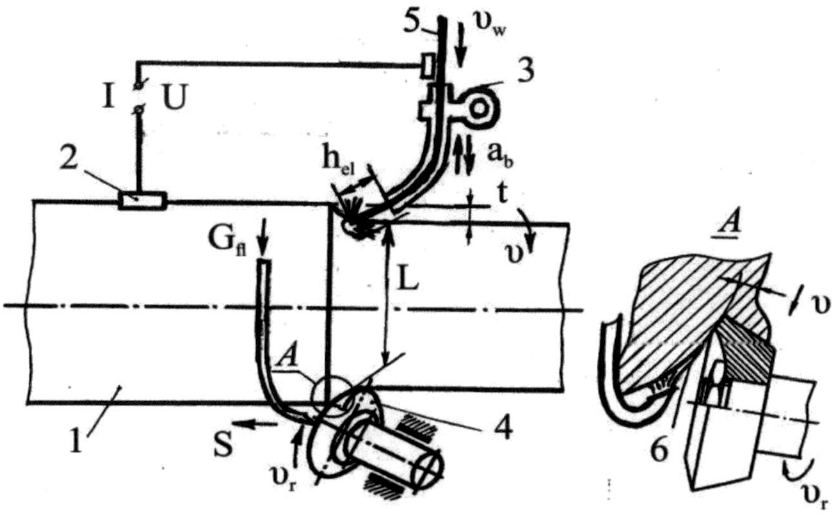

Figure 13.
Surfacing of wire with reinforcing rotational cutting: 1—workpiece; 2—sliding contact; 3—mouthpiece; 4—rotary cutter; 5—filler wire; 6—coolant; υ—main speed; υ_r—additional movement speed of the tool; S—feed rate; υ_w—wire feed speed; h_{el}—electrode overhang; a_b—vibration amplitude; G_{fl}—fluid flow; I—current strength; U—arc voltage; t—cutting depth; L—distance from electrode to cutter.

The study of the camshafts recovery showed that the surface hardness stabilizes (fluctuations within 3 ... 6 HRC) during electric arc surfacing of NP-30HGSA (the chemical composition of the wire %: C 0.25–0.35; Mn 0.8–1.2; Cr 0.8–1.2; Ni $\leq$ 0.40; Si 0.8–1.2; P $\leq$ 0.025; S $\leq$ 0.025) wire in a CO_2 environment, while the original parts had a significant variation (up to 30 HRC). After surfacing the wire, the hardness set in the technical documentation is ensured by subsequent heat treatment.

To eliminate the scatter of quality parameters of worn surfaces on the hardness of the surface layer and to ensure the physicomechanical properties of the layer located under the weld wire, it is recommended to use electromagnetic cladding (**Figure 14**).

When wear of the bearing journals surfaces exceeds the allowable limits, surfacing with wire is carried out on the insufficiently tempered surface. As a result, the hardened layer is formed on the surface basis not solid enough, which leads to the camshaft distortion. Preliminary hardening of the surface with ferromagnetic powder allows doping both the base and the surfaces formed during wire hardening (**Figure 15**) [18, 24, 25].

The surface geometrical parameters (radial runout ρ, dimensional accuracy IT, surface roughness Ra) after roughing are inherited in the finishing operations of grinding the bearing journals and cams of the camshaft. The geometrical deviations of surfaces after editing are saved on subsequent machining and assembly operations.

The analysis of the dependences of the influence of technological factors on the heredity of quality parameters made it possible to identify the determining processes of transferring properties when recovering, strengthening, and processing worn surfaces of the bearing journals and cams of the UMZ-4173 engine camshaft.

The analysis results showed that in the processes of electromagnetic surfacing of ferropowders and subsequent electric arc surfacing of NP-30HGSA wire on the bearing journals, as well as the arc current strength, magnetic induction, feed rates, and main processing movement affect the surface hardness in the process of plasma metallization of the cams using PG-10 N-01 (nickel-chrome base) powder. The determining parameter for quality control in the surfacing processes is the current strength.

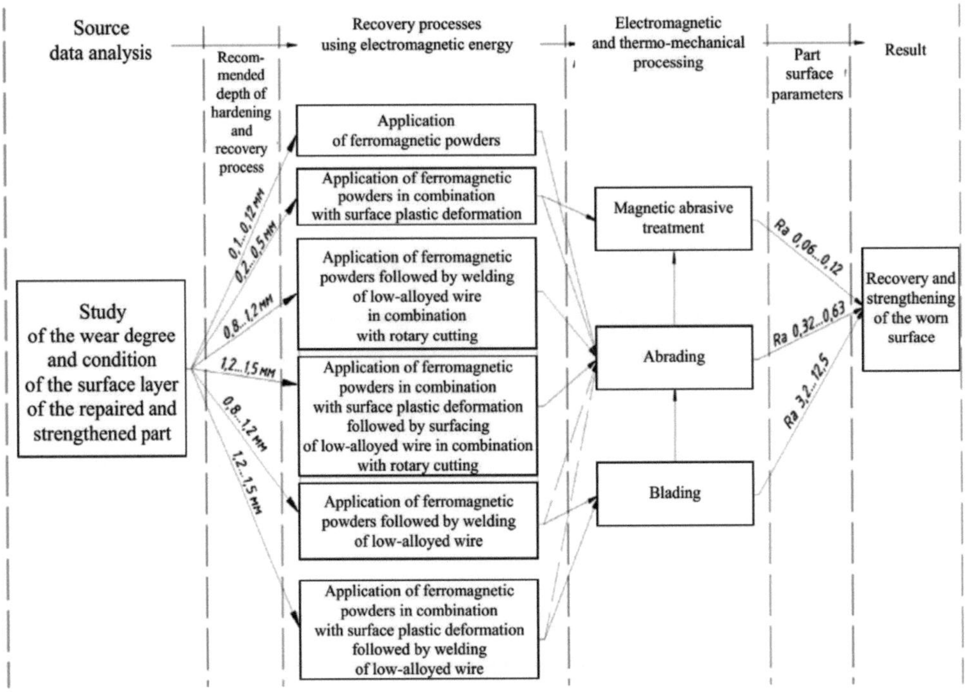

Figure 14.
The diagram of the recovery of the working surfaces of parts with varying wear degrees.

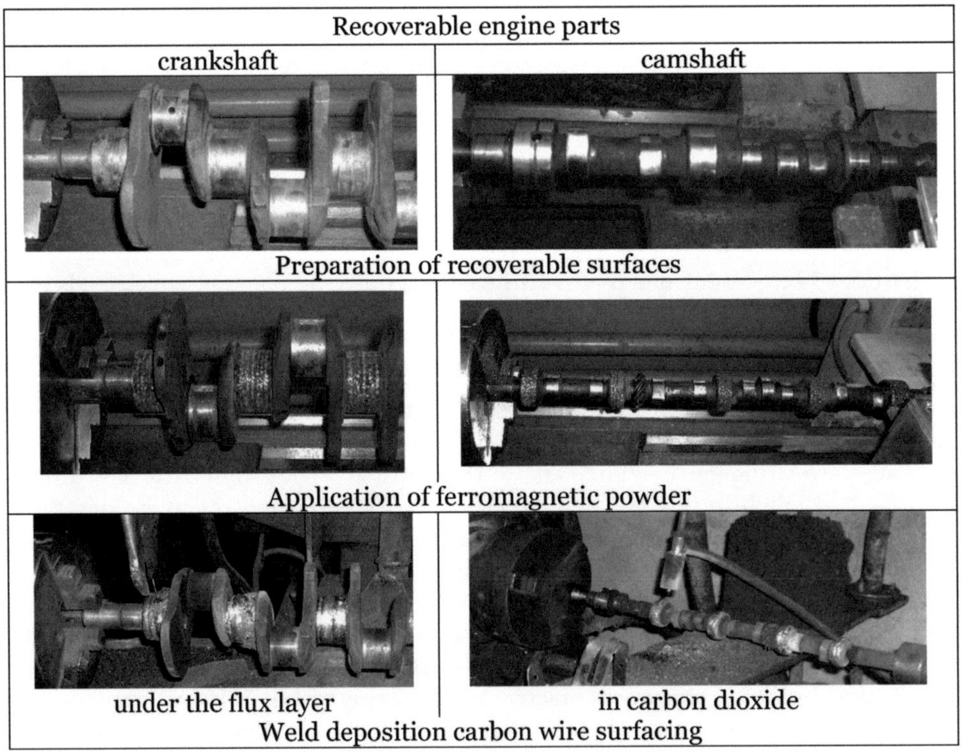

Figure 15.
Combined parts recovery technology.

In the final grinding, the hardness HRC and the roughness Ra of the surface are influenced by the radial and tangential components of the cutting force, which are determined by the depth of cut and feed during grinding, as well as the wheel and workpiece rotation rates.

Therefore, in controlling the machining quality, the focus should be on the depth of cut and the feed of the grinding wheel. The use of magnetic abrasive machining of polished profile surfaces can significantly reduce the duration of rolling operations.

According to the studies on the route of recovery operations, it was recommended: to provide stable hardness and uniformity of the coating material in the process of surfacing, and high surface hardness (54 ... 56 HRC) in the process of tempering; to eliminate straightening operations when recovering the camshaft to reduce the mutual radial runout of the surfaces to 0.02 mm; and to ensure the required accuracy of working surfaces.

The conducted research allowed to identify the processes of transferring the properties during recovery, hardening, and surface treatment with varying wear degrees of the bearing journals and cams of the UMZ-4173 engine camshaft, and to develop regulations for the technological process operations in accordance with them [26, 27].

6. Conclusions

The technological heredity of operational properties in the processes of manufacturing machine parts is advisable to describe by the graph reflecting the coefficients of transmission and mutual influence of physicomechanical and geometric parameters. To calculate the heredity coefficients according to the degree of the influence significance, a sequence of parameters is recommended: hardness, shape deviation, dimensional accuracy, and surface relief roughness of the part surface.

Methods of technological management and control of the heredity of operational properties of parts include: measurements of physical, mechanical, and geometric parameters of the most critical parts; determination of technological heredity mechanisms on the basis of transfer coefficient and coefficient of mutual influence of operational properties; analysis of technological barriers during intensive effects of energy flows; and development of measures for technological management of technological processes.

The technological heredity in the process of recovering the bearing journals and cams of the camshaft is non-monotonous and is fundamentally different from rational heredity with monotonous transfer of properties during machining, while 10–20% of indicators related to the shafts, the working surfaces of which are worn out with more than acceptable values, are out of the general dependence of the quality parameters transfer.

When recovering the surfaces, the geometrical and physicomechanical quality parameters of the camshafts first deteriorate and then improve; so the heredity is described by transfer and is determined by a uniform change in the hardness of the bearing journals and cams, and upon completion of the technological process, the geometrical characteristics are better than the original ones on worn surfaces, and the physicomechanical properties are recovered completely.

According to the study of the properties transfer processes when recovering the worn surfaces, it was recommended: to control the deformation of the part after machining operations; to eliminate the editing operation after heat treatment; to use a combination of methods and a combination of technological effects in recovering the parts surfaces with wear exceeding the limit values; to ensure stable physicomechanical properties of coating materials in the processes of surfacing and subsequent tempering; and to regulate the depth of cut and the supply of an abrasive wheel when grinding the recovered surfaces.

Thanks

Their research is supported by the Russian Fond Fundamental Research (Project 18-58-00009-Bel_a) and the Belorussian Republican Fond Fundamental Research (Project T18P-183).

Author details

Alexey G. Kolmakov[1], Mikhail L. Kheifetz[2,3*], Nikolay L. Gretskiy[3] and Gennady B. Prement[4]

1 A.A. Baikov Institute of Metallurgy and Material Science of the Russian Academy of Sciences, Moscow, Russia

2 Department of Physical and Technical Sciences of the National Academy of Sciences of Belarus, Minsk, Belarus

3 Open Joint Stock Company "NPO Center" National Academy of Sciences of Belarus, Minsk, Belarus

4 Open Joint Stock Company "Feloct-Service", Minsk, Belarus

*Address all correspondence to: mlk-z@mail.ru

References

[1] Kheifetz ML. The Design of Combined Machining Processes. Moscow: Mashinostroenie; 2005. 272 p. (Rus)

[2] Vasil'ev AS, Dal'skij AM, Klimenko SA, Polonskij LG, Kheifetz ML, Yashcheritsyn PI. Technological Bases of Quality Management for Machines. Moscow: Mashinostroenie; 2003. 256 p. (Rus)

[3] Yashcheritsyn PI, Ryzhov EV, Averchenkov VI. Technological Heredity in Mechanical Engineering. Minsk: Nauka i Tekhnika; 1977. 256 p. (Rus)

[4] Dal'skij AM. Technological Support for the Reliability of Highprecision Machine Parts. Moscow: Mashinostroenie; 1975. 223 p. (Rus)

[5] Yashcheritsyn PI, Skorynin YV. Efficiency of Friction Units of Machines. Minsk: Nauka i Teakhnika; 1984. 288 p. (Rus)

[6] Volovik EL. Handbook of Parts Recovery. Moscow: Kolos; 1981. 351 p. (Rus)

[7] Al'gin VB, Blyumenshtejn VY, Vasil'ev AS, et al. Technological and Operational Methods to Ensure the Quality of Machines. Minsk: Belaruskaya Navuka; 2010. 109 p. (Rus)

[8] Chizhik SA, Vityaz' PA, Kheifetz ML, et al. Quality Assurance of Products in Technological Complexes. Minsk: Belaruskaya Navuka; 2019. 248 p. (Rus)

[9] Haken H. Synergetics. Moscow: Mir; 1980. 404 p. (Rus)

[10] Yashcheritsyn PI, Averchenkov VI, Kheifetz ML, Kuhta SV. The analysis of the properties of technological solutions in the design of combined materials processing methods. Doklady of the National Academy of Sciences of Belarus. 2001;**45**(4):106-109. (Rus)

[11] Yashcheritsyn PI, Kheifetz ML, Klimenko SA, Vasil'ev AS. Technological and operational heredity of quality indicators in the life cycle of engineering products. Doklady of the National Academy of Sciences of Belarus. 2005;**49**(2):111-116. (Rus)

[12] Arzamasov BN, Krasheninnikov AI, Pastukhova ZP, Rakhshtadt AG. Scientific Fundamentals of Materials Science. Moscow: NE Bauman Moscow State Technical University; 1994. 366 p. (Rus)

[13] Gulyaev AP. Thermal Steel Treatment. Moscow: Mashgiz; 1960. 495 p. (Rus)

[14] Bochvar AA. Fundamentals of Thermal Processing of Alloys. Moscow-Leningrad: Metallurgizdat; 1940. 298 p. (Rus)

[15] Kolesnikov KS, Balandin GF, Dal'skij AM. Technological Basis for Ensuring the Quality of Machines. Moscow: Mashinostroenie; 1990. 256 p. (Rus)

[16] Dal'skij AM, Bazrov BM, Vasil'ev AS, et al. Technological Heredity in Engineering Production. Moscow: Moscow Aviation Institute; 2000. 364 p. (Rus)

[17] Yashcheritsyn PI, Kheifetz ML, Tochilo VS, Kusakin NA. The control of special processes of integrated and combined processing of machine parts. Doklady of the National Academy of Sciences of Belarus. 2004;**48**(6):113-118. (Rus)

[18] Yashcheritsyn PI, Kusakin NA, Kheifetz ML, Prement GB. Technological and operational heredity of quality indicators in the restoration of worn surfaces of parts. Doklady of the

National Academy of Sciences of Belarus. 2007;**51**(6):121-126. (Rus)

[19] Kusakin NA, Tochilo VS, Kheifetz ML. Quality Management of Autotractorrepair Company. Novopolotsk: Polotsk State University; 2009. 180 p. (Rus)

[20] Kheifetz ML, Gretsky NL, Kozhuro LM. Electromechanics of impulse surfacing processes of coatings by ferromagnetic powders. Strengthening Technologies and Coatings. 2008;**28**:51-57. (Rus)

[21] Kheifetz ML, Gretsky NL, Kozhuro LM. Thermophysics of surfacing processes of coatings by ferromagnetic powders in the electromagnetic field. Strengthening Technologies and Coatings. 2008;**30**:3-8. (Rus)

[22] Groche P, Bruder E, Gramlich S, editors. Manufacturing Integrated Design: Sheet Metal Product and Process Innovation. NY: Springer; 2017. 336 p. DOI: 10.1007/978-3-319-52377-4

[23] Youssef HA, El-Hofy HA, Ahmed MH. Manufacturing Technology: Materials, Processes, and Equipment. NY: CRC Press; 2017. 861 p. DOI: 10.1007/978-1-138-07213-8

[24] Kheifetz ML, Lysov AA, Kusakin NA, Prement GB. Controlling the heredity of quality indicators in the restoration, processing, assembly and operation of machine parts. Assembling in Mechanical Engineering and Instrument-Making. 2009;**105**:30-35. (Rus)

[25] Vasil'ev AS, Kondakov AI, Klimenko SA, Kheifetz ML, Gajko VA. Technological control of the heredity of operational quality indicators of hardened surfaces. Strengthening Technologies and Coatings. 2011;**62**:32-38. (Rus)

[26] Kheifetz ML, Prement GB, Gretsky NL. Technological control of the heredity of operational quality parameters in the engine camshaft recovery. Advanced Materials & Technologies. 2019;**1**:12-20. DOI: 10.17277/amt.2019.01.pp.012-020

[27] Kheifetz ML, Vasil'ev AS, Klimenko SA. Technological control of the heredity of operational quality parameters for machine parts. Advanced Materials & Technologies. 2019;**2**:8-18. DOI: 10.17277/amt.2019.02.pp.008-018

Chapter 9

Novel Antifouling and Self-Healing Eco-Friendly Coatings for Marine Applications Enhancing the Performance of Commercial Marine Paints

George Kordas

Abstract

Nanocontainers of the type CuO and CeMo were filled with bromosphaerol and 8-hydroxyquinoline, respectively, and incorporated into commercial marine paints. The generated paints with nanotechnology perform better in laboratory tests with respect to fouling and corrosion and test carried out via painting commercial ship traveling across the Adriatic Sea for a year than the currently used commercial paints. This is another application of nanotechnology that will someday find a commercial application. Since copper oxide is used for the current commercial paints and bromosphaerol is a natural biocide, there will be no need to pass the expensive approvals to use these antifoulants.

Keywords: nanocontainers, antifouling, marine paints, natural biocides, corrosion

1. Introduction

Biofouling is produced by immersing a surface (e.g., metal, net, etc.) in seawater where quickly a biofilm is formed first, composed of microorganisms, and then colonized by invertebrate animals. Next, corrosion is developed consisting of iron oxide, oxyhydroxides, green rust, etc. The metabolic activity influences further the corrosion process. All this generates problems to the structures exposed to the seawater with considerable consequences to energy consumption, speed of the ship, pollution of the environment, intervals between repairs, etc. [1]. This can be prevented using antifouling coatings composed of biocides and copper oxide preventing the organisms of settling. The total concentration of copper oxide in paints varies between 20 and 76 wt.% [2]. The biocides target microorganisms that produce a biofilm (typical bacteria). Biofouling is also a problem for all aquaculture industries, membrane bioreactors, offshore oil platforms, desalination units, cooling water systems, wind farms, oil pipelines, etc. Concerning antifoulants, tributyltin compounds (TBT $(C_4H_9)_3Sn$) were used for a long time and incorporated into marine paints [3]. It was determined that TBT was very toxic and prohibited to be used as antifoulant in 2008. This generated the demand to find natural antifoulants that are eco-friendly [4]. A number of marine metabolites show high-level antimacrofouling properties and were studied as additives for antifouling paints [5]. One of them, bromosphaerol, was identified that occurs in the red algae *Sphaerococcus coronopifolius*, inhibiting settlements of barnacles on marine paints [6]. Another strategy is to develop surfaces resembling that of shark because its surface is free of fouling [7].

It has been demonstrated that incorporation of nanocontainers loaded with inhibitors into coatings induces self-healing of paints after an external damage [8]. Among the various nanocontainer types, TiO_2 and CeMo exhibit also antimicrobial properties.

The present research was inspired by the previous successful use of nanocontainers loaded with inhibitors into paints (e.g., automobile and airplane multi-level protection of materials for vehicles by "smart" nanocontainers [9]) to tackle the replacement of chromium (VI) that is also carcinogenic. Since copper oxide is heavily used in marine paints as antifoulant, this inspired the production of copper oxide nanocontainers that were filled with bromosphaerol and SeaNine™211. The results of the laboratory work were encouraging and certified via painting two ships, the one traveling in the Adriatic Sea and the other in the Nord Sea for 1 year.

2. Experimental procedures

2.1 Experimental techniques

The nanocontainers were observed by scanning electron microscopy (SEM) and transmission electron microscopy (TEM). The corrosion behavior of the samples was determined using a Solartron ModuLab XM MTS equipment. The size of the nanocontainers was determined by a Malvern Nano Series system.

2.2 Chemicals

We used cerium (III) acetylacetonate (Sigma-Aldrich Chemie GmbH), sodium molybdate (Panreac Quimica SA), copper (II) sulfate pentahydrate (Sigma-Aldrich Chemie GmbH), polyvinylpyrrolidone (PVP, Sigma-Aldrich Chemie GmbH), potassium persulfate (KPS, Sigma-Aldrich Chemie GmbH), sodium dodecyl sulfate (SDS, Sigma-Aldrich Chemie GmbH), sodium chloride (NaCl, Sigma-Aldrich Chemie GmbH), 8-HQ (Sigma-Aldrich Chemie GmbH), bromosphaerol (FORKYS AE Ichthyokalliergeies) and 1-BSA (Sigma-Aldrich Chemie GmbH).

2.3 Nanocontainer production

The nanocontainer reduction has been the subject of many of our papers. We sketch here the production of CeMo and CuO that we cannot claim an extensive description and advise the reader to search for more details in our literature [8].

2.3.1 CeMo nanocontainers

The CeMo nanocontainers were produced using a three-step process. The first process involves the formation of a core. This was accomplished by mixing 10.0 g styrene, 10.0 g potassium persulfate, 1.3 g sodium dodecyl sulfate in 900 g water in a 500 cm^3 container. The flask was purged under nitrogen for 18 h. The second step was the coating of the polystyrene core by $Ce(acac)_3$ and sodium molybdate aqueous solution in the presence of PVP. The third step was the drying of the powder at 60°C for 1 h and then heating at 550°C for 4 h. The process was described extensively in the literature [8].

2.3.2 CuO nanocontainers

0.75 g of $CuSO_4 \cdot 5H_2O$ and 0.2 g of glucose were diluted in 0.2 L of distilled water. Then, 50 mL of $NH_3 \cdot H_2O$ solution (0.04 M) and 50 mL of NaOH solution (0.20 M) were assorted slowly in 30 min. At the end, 0.100 mL of ascorbic acid (0.03 M) was added. After this, the color of the mixture from blue turned into yellow. Later, the color changed into orange. We centrifuged the solution at 8000 rpm for 5 min to receive the solids. We processed Cu_2O at 250°C for 2 h to observe a black solid of CuO hollow nanospheres via thermal oxidation. The hollow CuO nanocontainers were loaded via a vacuum facility with bromosphaerol.

2.3.3 Marine paints

We used two commercial paints for the experiments from Wilckens. The paints were free from any additives used for corrosion and antifouling protection (**Figure 1**).

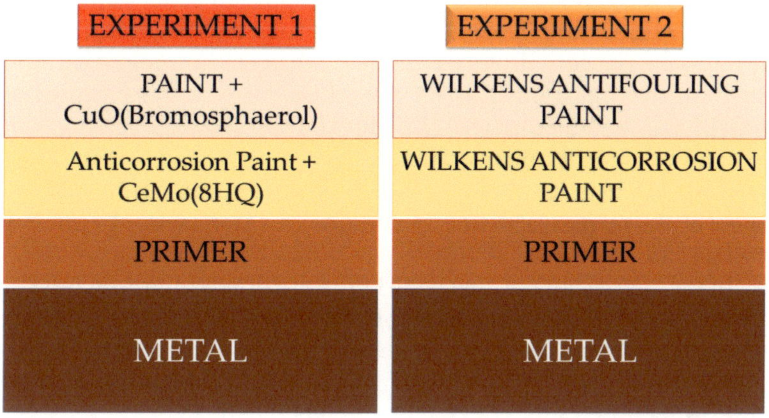

Figure 1.
Two experiments where metals were painted: the first with the primer (bottom), anticorrosion paint with CeMo(8HQ) (mittle), and paint with CuO(bromosphaerol) (top) (Experiment 1); and the second the primer (bottom), anticorrosion Wilckens paint (mittle), and paint with Wilckens antifouling paint (top) (Experiment 2).

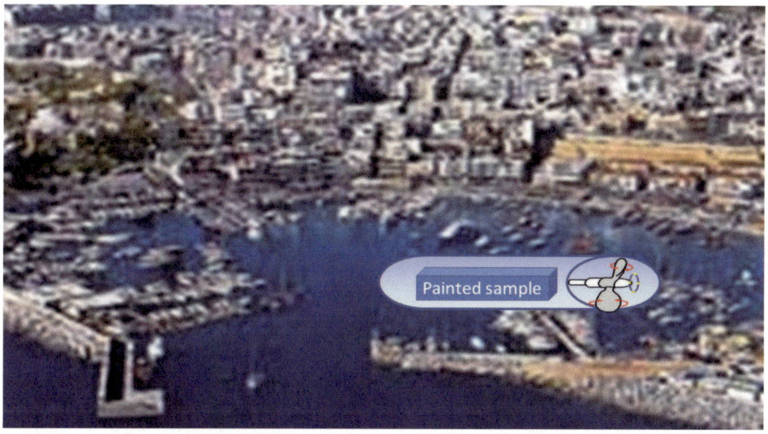

Figure 2.
Tube with samples placed in Mikrolimano (Piraeus) for 3 months. The motion of a ship with 14 knots simulated using at the end of the tube a propeller.

2.3.4 FRA measurements

Metals used for ship hulls were painted in size 1 cm by 2 cm and were placed in a tube in the seawater in the Mikrolimano harbor (**Figure 2**). At the end of the tube, a propeller was placed inducing motion of water inside the tube corresponding to 14 knots. After 3 months of exposure of the samples in this custom-made facility, corrosion tests were conducted using the electrochemical impedance spectroscopy (EIS).

3. Results

Figure 3 shows the SEM micrographs of the CuO and CeMo nanoparticles. One can perceive a size of ~420 and ~350 nm for the CuO and CeMo nanocontainers. The term "nanocontainers" is used instead of "nanoparticles" because the nanomaterial produced here is hollow inside as one can perceive from **Figure 2**.

The CuO and CeMo nanocontainers were filled with bromosphaerol and 8-hydroxyquinoline, respectively. This was accomplished using a special for this purpose vacuum facility. The question is how much of bromosphaerol and 8-hydroxyquinoline the containers were filled. We used thermogravimetric analysis (TGA) to answer that question. **Table 1** gives the results for the encapsulation efficiency (EE) and loading capacity (LC) for the two systems.

Figure 4 shows the SEM micrographs of the three coatings: (a) the primer, (b) primer plus anticorrosion, and (c) the primer, anticorrosion, and antifouling layers. The thickness of the primer is 50 μm, of the anticorrosion layer 140 μm, and of the antifouling layer 130 μm.

The samples were placed for 3 months in the facility in Mikrolimano (Piraeus) and then soaked in 0.5 M NaCl solution for 48 h. The FRA spectra of the two samples the one painted with our technology (Experiment 1) and the other painted with the commercial paints (Wilckens Experiment 2) are shown in **Figure 5**. One can perceive from **Figure 5** that the nanotechnology paints are more stable in the seawater for 3 months than the commercial paint.

The nanotechnology paints were tested using one passenger ship (Sea Anemos) traveling daily between Ancona (Italy) and Patras (Greece). **Figure 6** shows the

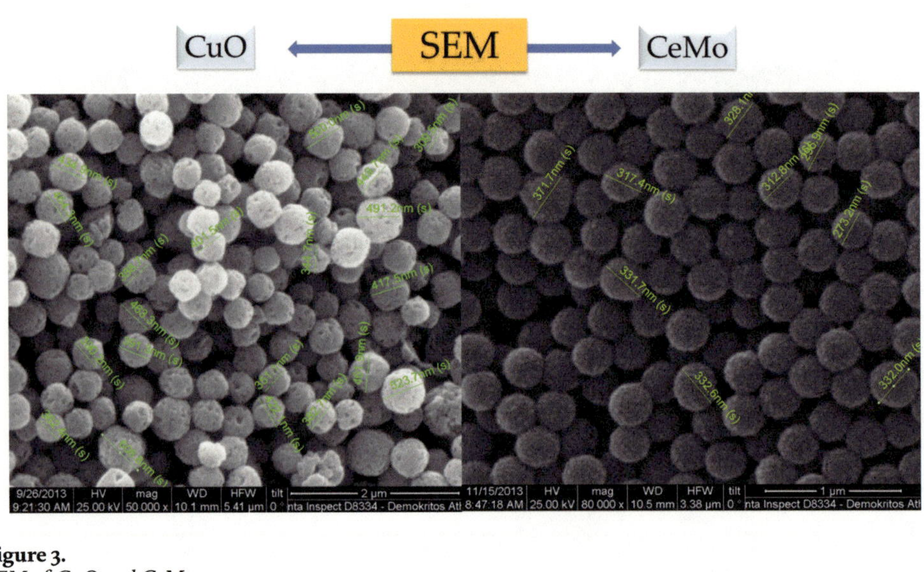

Figure 3.
SEM of CuO and CeMo.

	EE	LC
Bromosphaerol in CuO	50%	25%
8-OH quinoline in CeMo	80%	40%

Table 1.
EE and LC for bromosphaerol and 8-OH quinoline in CuO and CeMo, respectively.

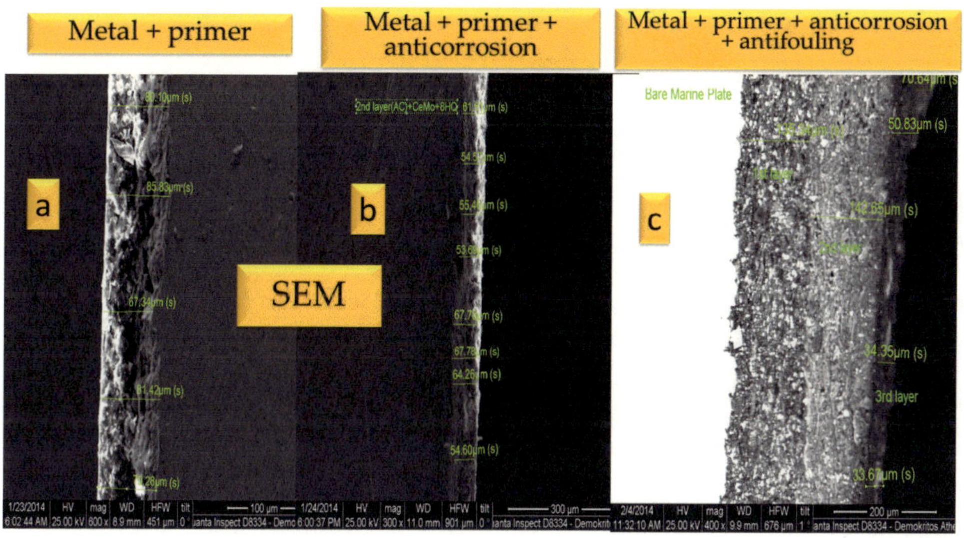

Figure 4.
SEM of primer, anticorrosion, and antifouling layers.

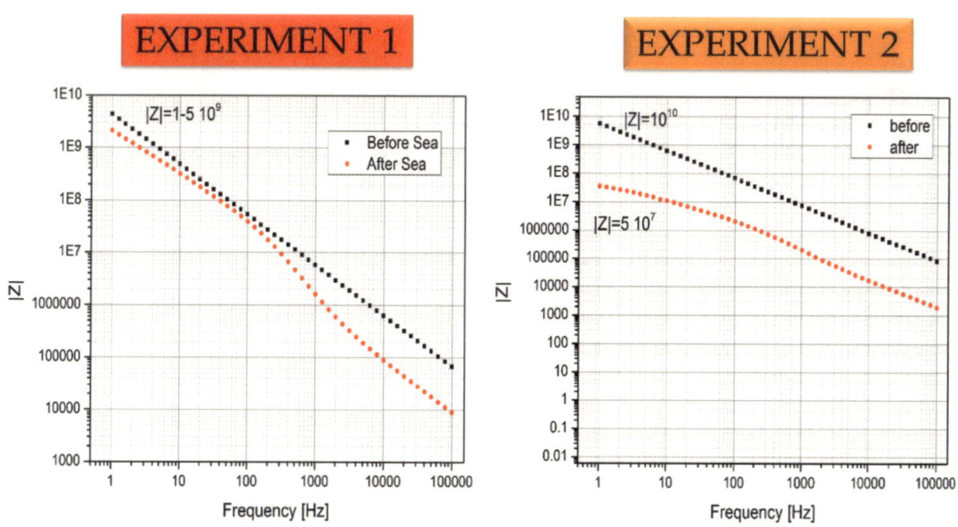

Figure 5.
FRA of the nanotechnology-based coating (Experiment 1) and of the commercial paints (Experiment 2).

stripe that was painted, and the rest of the ship was painted with the commercial coatings (primer, anticorrosion, and antifouling paints). After a year, the ship was taken out in a shipyard near Piraeus (Skaramagas shipyard, in a port town in the western part of Athens), and the result is shown in **Figure 6**. To our surprise, the

Figure 6.
Ship painted with nanotechnology.

stripe we painted with the "nanotechnology paints" was nearly without fouling as compared with the rest of the ship painted with the "commercial paint."

4. Discussion

In the last few years, polymer-based nanocomposites incorporating nano-sized fillers have been studied extensively. The merits of this approach are that the nano-sized particles provide a larger interface for interactions with polymer matrix leading to improved properties. The control of the surface topography at nano level leads to new types of coatings with targeted properties, like antifouling. It is important to mention here the "nanostructuring" of surfaces can alter the wetting properties of the surfaces, the so-called lotus effect, leading to self-cleaning or super hydrophobic coatings. The presented biomimetically inspired surface topography Sharklet AF™ containing 2 μm wide rectangular-like periodic features spaces at 2 pm that reduced Ulva settlement by 86%. This is a convincing demonstration that antifouling can be achieved by the topography of the paint surfaces inhibiting accumulation of marine algae. When spacing above 5 μm this feature, the inhibitory response is not achieved. There is under development a new class of antifouling coatings called "biomimetic" resulted from the skin of shark consisting of overlapping plates in nanoscale size with parallel-oriented ridges preventing shark from becoming fouled even standing. Many investigators were inspired by this fact that is formulated by the engineering roughness index (ERI) given by Eq. (1):

$$\mathrm{ERI} = \frac{r * n}{1 - \varphi} \tag{1}$$

where r is the Wenzel roughness ratio, n is the number of individual surface features, and φ is the area segment of the dissimilar surface topographies. When the surface is completely smooth, ERI is zero. The Sharklet AF exhibits an ERI of 9.5 due to 3D dimensions in microstructural differences. Due to this patterning, Sharklet AF conveys a 77% decrease in microfouling.

Considering the fact that the nanostructure impacts the topography of the surfaces, one can use these observations to explain the outcome of the paintings of the ship with and without nanocontainers on the surface shown in **Figure 6**. Another contribution arises from bromosphaerol included in the nanocontainers given in **Table 1**. It has been demonstrated that bromosphaerol inhibits settlements of barnacles on painted substrates [6]. Bromosphaerol is released from the nanocontainers via their porosity shown in **Figure 3** or via wearing out the nanocontainers by the friction with the seawater while the ship is moving in the sea. It is known that copper oxide is also an antifoulant. The commercial paints include copper oxide in the amounts of 20–76 wt.% in the paints. Here, we used 4 wt.% CuO in the form of nanocontainers. Though the amount is small, one has not to neglect the antimicrobial properties observed in nanocontainers [10]. All these factors together may lead to the outstanding outcome of **Figure 6**. Here, the contribution arises not only by one property, namely, the shark skin effect, but from the contribution of the bromosphaerol that is a natural occurring diterpene from the red algae *Sphaerococcus coronopifolius,* which can inhibit settlement of barnacles on painted substrates and the antimicrobial contribution of the CuO nanocontainers. All three components lead to a superb technology not to be achieved by other methods.

The nanotechnology strategy of this work has also another implication regarding the anticorrosion properties of the paints. Experiments 1 and 2 of **Figure 5** show the response of the paints including CeMo(8HQ) into the anticorrosion paints (Experiment 1) in contrast to the commercial paints of Wilckens (Experiment 2). One can perceive from **Figure 5** a Rp of 10^{10} Ω for the as painted substrate (Experiment 1), and it remains about the same after exposure of the samples for 3 months into the seawater in the Mikrolimano. On the contrary, the commercial paint has Rp equal to 10^{10} Ω before immersing the sample into the seawater and is reduced almost three orders of magnitude after 3 months in the seawater, the Rp equal to 5×10^7 Ω. Again, the nanotechnology-produced samples exhibit superior properties. The difference between the value of Rp of the commercial paint and the Rp of the paint with CeMo(8HQ) can be attributing to self-healing effect offered by the CeMo(8HQ) nanocontainers. The new technology is dynamic instead of static paints. The result of the painted ship tells us that the technology is not good for the laboratory but ready to be exploited with many implications not only for the ship hulls but also for aquaculture nets, oil drilling platforms, pipelines, etc.

Nanotechnology will contribute to many areas with outstanding impact in nanomedicine, energy, energy storage, thermal energy conversion, etc. Here, we demonstrated a contribution in the field of preventing biofouling. If one considers the impact to economy, air pollution reduction, increase of the time between repairs of the ship hulls, reduction of cost of paints due to the reduction of the use of copper oxide, increase of speed of travel due to the increase of the contact angle of seawater reducing the drag resistance, and many other benefits, we are experiencing a new revolution in technology arising from nanotechnology. Nanotechnology will not be like other technologies where money was wasted experiencing a boom with many expectations. Any investment in nanotechnology science and technology will deliver a multiple return to the investment with results benefiting the societies like cancer therapy, energy storage, self-healing of materials, etc.

5. Conclusions

The present technology will compete with existing technologies of reputable maritime companies. Sigma developed the product SYLADVACE™ 700 built upon pure silyl acrylate binder. This product exhibits self-smoothing characteristics of the silyl acrylate binder reducing the leached layer and average hull roughness. International Paint Ltd. (AkzoNobel) developed the Intersleek™900 product, yielding a smooth, slippery, low-friction surface onto which fouling organisms attach very difficult. Chugoku Marine Paints, Ltd. has the Seaflo Neo™ SL products in the market attaining an ultrasmooth surface with good self-leveling performance.

HEMPEL established a self-polishing antifouling paint named Clobic™ NCT based on nanocapsule technology. HEMPEL has also the HEMPASIL™ X3 87,500 with high solid content. JOTUN has the SeaQuantumPlus™ S in the market for faster vessels. NIPPON has launched the A- LF-Sea™. By incorporating this into NIPPON PAINT'S excellent self-polishing copper silyl acrylate antifouling, long-term low friction performance is ensured, and 10% more reduction of fuel consumption is guaranteed.

The nanotechnology developed in the Sol–Gel laboratory of NCSR "D" will not be able to break through these technologies of the big companies in the field. This work offers a brainstorming approach that these companies consider such an interesting approach because this technology offers self-healing technology together with excellent antifouling properties. This new technology will not need permissions to use in various countries as we all know how much time consuming they are and also expensive. The present work uses natural products as biocides and nanocontainers like CuO that is used to an amount of 4 wt.% instead of 20–76 wt.% of the current technology. Our technology uses less amount of copper oxide reducing significantly the cost of the paints.

Acknowledgements

The work is financed by the Greek Secretary for Research and Technology under contract ΕΣΠΑ 1274.

Author details

George Kordas
Sol-Gel Laboratory, INN, NCSR Demokritos, A. Paraskevi Attikis, Greece

*Address all correspondence to: gckordas@gmail.com

References

[1] Aldred N, Clare AS. The adhesive strategies of cyprids and development of barnacle-resistant marine coatings. Biofouling. 2008;**24**(5):351-363. DOI: 10.1080/08927010802256117

[2] Hellio C, Yebra D, editors. Advances in Marine Antifouling Coatings and Technologies. Boca Raton, FL: Woodhead/CRC; 2009. ISBN: 9781601199997 (Electronic book)

[3] Armstrong E, Boyd KG, Burgess JG. Prevention of marine biofouling using natural compounds from marine organisms. Biotechnology Annual Review. 2000;**6**:221-241. DOI: 10.1016/S1387-2656(00)06024-5

[4] Schultz MP. Effects of coating roughness and biofouling on ship resistance and powering. Biofouling. 2007;**23**:331. DOI: 10.1080/08927010701461974

[5] Townsin RL. The ship hull fouling penalty. Biofouling. 2003;**19**:9-15. DOI: 10.1080/0892701031000088535

[6] Fusetani N. Antifouling marine natural products. Natural Product Reports. 2010;**28**:400. DOI: 10.1039/c0np00034e

[7] Li G. Preparation, anti-biofouling and drag-reduction properties of a biomimetic shark skin surface. Biology. 2016;**5**:4. DOI: 10.1242/bio.016899

[8] Montemor MF, Snihirova DV, Taryba MG, Lamaka SV, Kartsonakis IA, Balaskas AC, et al. Evaluation of self-healing ability in protective coatings modified with combinations of layered double hydroxides and cerium molibdate nanocontainers filled with corrosion inhibitors. Electrochimica Acta. 2012;**60**:31-40. DOI: 10.1016/j.electacta.2011.10.078

[9] Available from: https://cordis.europa.eu/project/rcn/89916/factsheet/en

[10] Trapalis CC, Kokkoris M, Perdikakis G, Kordas G Study of Antibacterial Composite Cu/SiO2 Thin Coatings. Journal of sol-gel science and technology, 2003;**26**(1-3):1213-1218

Chapter 10

Durability Issue for the Emperor Fasiladas Royal Palace in Gondar (Ethiopia)

Eskinder Desta Shumuye and Gashaw Assefa Bezabih

Abstract

The Royal Enclosure is the remains of a fortress-city in Gondar, Ethiopia. It was founded in the 17th century by Emperor Fasiladas and was the home of Ethiopia's emperors. Its unique architecture shows diverse influences including Nubian styles. The site was inscribed as a UNESCO World Heritage Site in 1979. Ghebbi is an Amharic word for a compound or enclosure. Due to climate conditions and human activities, the Royal palace is affected by severe structural damage. Presently almost some portion of the palace are under maintenance by mortar pointing to avoid negative effects of rainfall and other durability issue and temporary scaffolding to prevent from collapse of vulnerable structures. An analysis of damage of the palace is presented, based on weathering processes and structural conditions, as preliminary tool to detect and implement urgent and medium/long-term protection strategies for the conservation of the monuments. The chapter describes the major durability issue of the historical palace and determines the cause of the present durability problem and then recommends the possible remedial measure to alleviate the prolonged durability issue. The analysis was conducted by visual inspection and X-ray diffraction characterization methods. The chapter discusses the results obtained from the analysis of the mortar sample of the historical palace.

Keywords: durability, Gondar, Fasil Ghebbi palace, historical heritages, mortar

1. Introduction

Surrounded by a 900-metre-long wall with 12 entrances and three bridges Fasil Ghebbi resides on an area of about 70,000 square meters, Fasil Ghebbi is located in Gondar City in north western Ethiopia. The compound was Registered on the World Heritage List in 1979 and it contains palaces, churches, monasteries and public and private buildings. It served as the home of Ethiopian's emperor Fasiladas and various Emperors who ruled during this period in the 17th and 18th centuries. This is only one of its kind architectures shows diverse influences including Nubian, Arab, and Baroque styles.

Various rebuilding and repairs were undertaken from the late 1930s to the late 1960s. Under the Italian occupation (1936 to 1941), extensive and radical restoration work was done on many of the monuments, using cement and reinforced concrete [1, p. 79]. Due to Earlier inappropriate repairs together with changes in function have resulted in serious damage and major restoration work was carried out on the compound Under the UNESCO/UNDP assisted project ETH/72/014. The

Figure 1.
Fasil Ghebbi palace [29].

situation was partially reversed with the restoration works carried out by UNESCO in the 1970s, which replaced the cement and concrete work with the original mixes of lime mortar as well as with subsequent major conservation programs implemented since 1990 [2]. Even with the Mentioned Restoration efforts in 2013 Fasil Ghebbi was listed as one of the World Heritage in Danger by the united nations educational, scientific and cultural organization, and according to Gebrehiwot [3], recently, there were same activities by the Government of Ethiopia with the support of donors to conduct additional restoration and renovation works.

According to Eskinder [4] method of construction for the Royal palace walls were by stone masonry. The abundant type of construction materials which were used during the construction of the Royal Palace was mainly by Basaltic Rock and Pumice with a Local name of "Beha Dingay [ባሃ ድንጋይ]"; which is transported from "Kuskuam mariya" and "Azezo," respectively. The type of binding material which was used during the construction of the Royal palace was also identified as lime (CaO). It is clear that lime and gypsum have always been functionally important materials in building, and in the light of recent archeological investigation, it appears that these materials are equally important historically. And they are important not only in building history but in the general history of mankind [5] (**Figure 1**).

2. Methodology

To study the mineralogical composition of both the original binding mortar and the mortar used for restoration works, XRF (X-ray diffraction) technique were used to identify the change in mineralogical composition of binding materials. After extracting the mortar sample from the exterior wall side, mortar sample were crashed to a powdered and sieved from number 200 sieve [6, 7]. Hydration products such as $Ca(OH)_2$, $CaCO_3$, C-S-H, calcium silicate and quartz were also analyzed. Result from the XRF test of mortar sample also used to identify the difference in mineralogical composition between exposed mortar with that of the unexposed one.

Site visit also carried out for assessing the physical impact due to weathering problem, tilting of wall and mechanical impact related problem, presented using photographs. Pictorial comparison of the current situation of the palace building and appearance of the palace, 5 years before are also presented in this document to evaluate the impact of durability on the palace building. Furthermore, Hydrology and geological investigation are also covered.

3. Case of deterioration

Considering the importance of preserving historical buildings, identifying the method of construction, investigating the cause underlying the occurrence of degradation and establishing the degradation mechanism is a priority. Historical mortars are complex systems, containing aerial or hydraulic binders or blend of them, aggregates sometimes reactive siliceous minerals that interact with the alkaline binder and, often, some additions. The reactions among the constituents and with the environment go on for long time.

Preservation and restoration of historic buildings represent, worldwide, serious concerns within the cultural community. Many valuable historical buildings are degrading because of a complex and combined effect of decay process, such as climate causes, structural issue, biological and botanical issue, insect and animal causes, air pollution, poor construction and inappropriate repair are considered as the major causes of deterioration [8]. The evaluation of the influence of each of the environmental factors in a given situation requires an understanding of which mechanisms are potentially of concern for the type of material or structure in question. In general, the main deterioration mechanisms include: Moisture, Environmental impact and Chemical actions [9].

3.1 Moisture

Moisture and temperature affect the chemical, biological, and mechanical processes of decay. The formation of a moisture layer on the material surface is dependent upon precipitation. It may also be generated as a result of the reaction of adsorbed water with the material surface, deposited particles with the material surface, and deposited particles with reactive gases [9].

Diana-Andreea stated the source of moisture in a historical buildings as Infiltration of groundwaters, because of the capillary rise and of the forces of electro-osmosis, Defective collection of rainwater and the lack of an effective vertical systematization, Condensation of water vapors from the air and from the pores of the building materials [10].

The soil between groundwater level and earth surface holds water by capillarity where this event is called as "surface water" or "capillary water" that cannot be removed by using any drainage system. The humidity rising up the building can cause serious damages on the structure [11]. According [Nigussie] to The groundwater around Gondar city can be found at a depth between 6 and 8 meters [12] consequently the groundwater at this depth cannot affect the durability of the structure (**Figure 2**).

Köppen-Geiger climate classification system classifies its climate as humid subtropical (Cwa), bordering with subtropical highland (Cwb). The annual average rainfall of Gondar city, Ethiopia is 1151 mm [13]. Rainwater Does not directly affect the deterioration of the walls; however, it may be acting indirectly in the mechanisms of deterioration it dissolves the mortar and it has a leading role in the activation of the primary mechanisms of deterioration of limestone rocks [14]. The rain water and the presence of moisture due to the rain water damages the exterior portion of the palace. The rain water is a combination of other soluble materials that causes

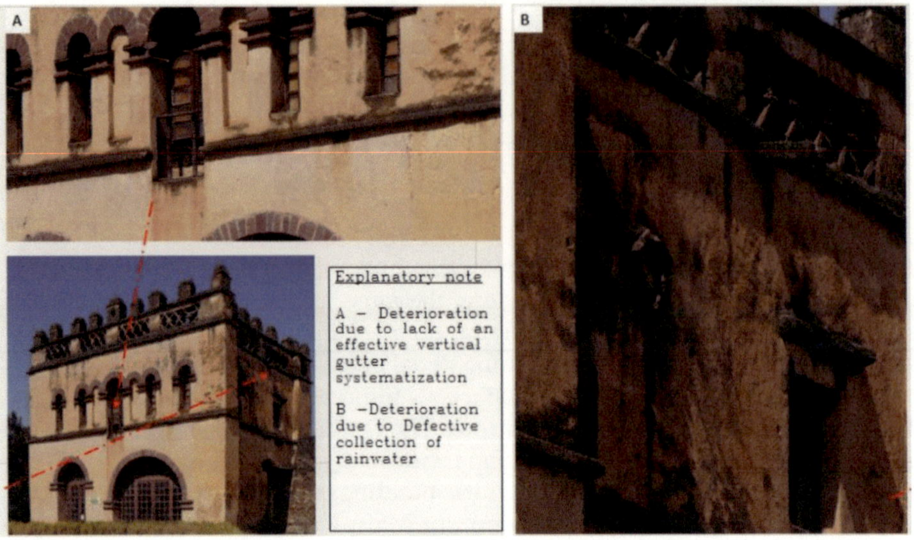

Figure 2.
The effect of improper rainwater removal on the face of the wall [29].

damaging manifestation when this rainwater vaporizes. In case of the Fasil castle, geomembrane was used to solve the effect of rain water dispersion through the wall cracks. However, the application is inadequate for the main palace. Two variables are important from the viewpoint of the damage caused by moisture: dew point and relative humidity of air.

Dew point is a characteristic of the water content of the large-scale air mass, whereas relative humidity depends on the local temperature and therefore on the local meteorological parameters. When the temperature of a material is below the ambient dew point, water condenses on the material, a moisture layer (condensation) can form, and the material damage may proceed. In most materials, an increase in relative humidity causes further deterioration due to more prolonged wetness time, higher deposition rates of pollutants and better conditions for biodeterioration [9].

Considering the fact that in order to properly develop heritage protection strategies, understanding of the future climatic conditions is of great importance, to come by/achieve these, various method have been developed in recent years in order to predict the evolution of certain meteorological parameters.

3.2 Chemical actions

Stone has been used as building materials for thousands of years. One of the reasons for this is the local availability of stones [15]. As it is proved in the Fasil Ghebbi palace, the appeal, stability and durability of stone are among the most important reasons for using it in stone masonry construction.

In addition to structural problems which is mainly causes by inappropriate methods of expatiation, defects due to poor workmanship and negative external effects [8]; atmospheric movement and humidity have played negative effects on conservation of historical buildings. Changes in temperature cause fragmentation and chipping in humid environments due to the presence of internal stress. Various acids formed by chemical reactions around plant roots lead to the deterioration of stones used during the palace construction. As seen in **Figure 3**, the roots of plants that flourish penetrate deep into the gaps between or the fractures on the stones used in the construction of historical buildings and induce chemical and physical weathering.

Figure 3.
The progressive biodeterioration processes on the surface of building materials: algae constitute the medium for the growth of fungi that pass through the building material by means of their hyphae. Dense formations of herbaceous plants [A] vascular plants [B] mosses and algae [source; Eskinder Desta].

Biological colonization is affected by substrate characteristics both physically and chemically. Due to the surface roughness of the masonry wall, algal cells can easily adhere to rougher surfaces, yet algae have the ability to extend on such surface. Microorganisms flourishing on the surface lead to microbiological deterioration and chemical deterioration. In most walls of the palace, which is exposed to the moisture; microorganism is observed on the facades of buildings, consist of bacteria and algae (including cyanobacteria), lichens and protozoa. Because of their resistance to high temperature and desiccation, lichens and vascular plants are generally play a role in the biodeterioration of the binding material of the masonry wall of the palace.

The formation of crusts tempted by cyanobacterial and algal growth also results in a longer moist retention at the wall surface, increasing the mechanical damage produced by the volume change in the pores of the stone [16] (**Table 1**).

The XRF result reveals that, from the total mineralogical composition of the original binding materials. The first abundant mineral from the original binding material is calcium which holds 18.7% of the mineralogical composition. The remaining mineralogical composition were occupied by quartz and rock fragment. From this result we can conclude that the amount of mineral composition which results cementitious properties of the binding materials is showing changes due the weathering effect.

All building materials are exposed to degradation processes which exert a stress on the building materials, then after a certain time it will cause major durability problem [17]. As seen from the XRF result the percentage of CaO is relatively less compared to the other cementitious materials, which results a lower cementitious property of the existing mortar. Degradation process is a gradual process leading to a reduction to the quality of the building materials.

Bio-deterioration of the binding material occurs due to alive organisms or the products of its metabolism. The process of colonization of a mortar is favored by its characteristics (porosity, mixture and roughness). These properties facilitate the retention of water in the binding material and the consequent growth of algae and cyanobacteria. As shown in **Figure 3**, it is the common durability issue in the royal

Oxide formula	Shorthand notation	Cement	Original binding material (lime) based
CaO	C	64.5	18.70
SiO2	S	17.945	35.46
Al2O3	A	4.4	7.98
Fe2O3	F	3.5	11.02
MgO	M	3.7	5.48
K2O + Na2O	K + N	1.2	2.65
Others	....	0.2664	2.22
H2O	H	—	5.12

Table 1.
Chemical composition of mortar samples from the main palace wall (%wt.) [4].

palace. For instance, biodeterioration of mortars of the Mosaics of Italica and some other Spanish archeological sites (Baelo Claudia, Medina Azahara) solved by the development of the biocide mortar, the biocide is adsorbed in sepiolite avoiding a quick leaching and increasing the effectiveness [18].

When lime is mixed with water, it forms calcium hydroxide, called slaked lime [19]:

$$CaO + H_2O \rightarrow Ca(OH)_2 \tag{1}$$

The reaction of calcium hydroxide with carbon dioxide is faster, forming mortar mixture that gain strength more quickly.

$$Ca(OH)_2 + CO_2 \rightarrow CaCO_3 + H_2O \tag{2}$$

Some cement compounds having both free alumina and silica, and the sulfates, forming sulfa-aluminates (ettringite) and sulfa-silicates, causes degradation of mortar. Mortars may suffer from external sulfate attack when exposed to sulfate-rich environments or from internal sulfate attack due to the presence of sulfate compounds in the constituents. Sulfate ions can react with the cement's hydration products to form ettringite and gypsum. These new phases encourage considerable expanding pressure of the adjacent cementitious matrix, resulting in cracking.

3.3 Maintenance issue

Conservation involves works undertaken to preserve the condition of the building to its original state and this also includes the subsequent maintenance works [20]. In most cases, having a well scheduled and properly constrictive maintenance is considered as a key factor for prolonging the life span of the historical structures. **Figures 3** and **4** reveals that, due to several factors a wide spread lack of binding mortar in between the stone masonry is noticed. To alleviate the above problems mortars and reduced the rock fragmentation; as shown in Figure mortar repointing has been submitted to may interventions in a certain period. Consequently, it is also challenging to determine the same kind of binding mortar as that of the one used in the constative stage. However, as Fernando et al. stated that during the rehabilitation of stone masonry walls, compatibility and durability of the intervention are very important [8, 21].

The major raw materials used for restoration works (lime) were transported from the south west part of Ethiopia, which makes it very challenging to perform a frequent restoration work. Some of the maintenance related issues of the Fasil Ghebbi palace are listed underneath [4, 22] (**Figure 5**) (**Table 2**).

Figure 4.
A widespread lack of binding mortar in between the stone masonry [source; Eskinder Desta].

Figure 5.
Mortar repointing used during the restoration of the palace [source; Eskinder Desta].

No.	Sample description	Lab. No.	TW/HWT (gm)	Magnetite	Ilmenite	Calcite	Quartz
01	Restoration site	20209/14	10/9.95	1	Tr	*50	2

*Remark: * = clay size calcite; R. F = rock fragment; Tr = trace element.*

Table 2.
Chemical composition of mortar samples from the restoration site [4].

3.3.1 The absence of guidelines regarding to maintenance for historic building

There are no specific guidelines, procedures, method and system established by the concerned authorities regarding to maintenance of historical buildings.

3.3.2 Unplanned approach

The unplanned approach is the main approach adopted in the most of historic building interviewed. In addition, most of the maintenance work only take an action whilst the elements of the historical heritage display failure and malfunction.

Figure 6.
Active durability issue due to the presence of grass and alga on the wood floor terrace [source; Eskinder Desta].

Figure 7.
Active durability issues risking the palace: (a) crack formation inside the palace internal partition wall (repointed surface), (b) steel frame support aimed for the enhancement of stone masonry dome structure, and (c) tilting of the stone masonry wall inside the main palace yard [source; Eskinder Desta].

3.3.3 Lack of maintenance staff training and expertise

The maintenance staff training, and expertise is the most important component in the maintenance process for historical building. Furthermore, the lack of maintenance staff training, and expertise also results for outsourcing their maintenance tasks to the contractor which even do not have related practical experience in the maintenance of historical heritages.

3.3.4 Lack of financial support

The main problem in maintenance of historical heritage building is also luck of financial support. The financial support is fundamental element in order to assurance the cultural significance will be preserve and conserve properly.

3.4 Visual inspection

Considering the stone masonry elements and their environmental conditions, the durability risks originates from both the aggressive agents in the atmosphere, human activity and seasonal biological colonization's. Based on the state-of-the-art knowledge, the possible deterioration process comes from different perspectives. Some of these buildings have deteriorated and still deteriorating for such reasons as intentional human activities, non-maintenance, climatic factors, and anthropogenic and biological impacts.

During the visual inspection of the main palace building, there were identified advanced signs of surface decay due to water infiltration in the basement and fissures on the wall and wooden floor of the main palace balcony. Following are some of the most problems spotted in the fail Ghebbi palace. Along with the respective reference figures.

a. Emission of vascular plant roots lead to the degradation of rocks and binding materials, which increase the formation of cracks in between the masonry stone layers and wall surface and thus facilitating deterioration.

b. The presence of vascular plant affects the visual esthetics of building by covering the surface of the buildings.

c. Seeds and small tree located on the building attracts animals and insects such as birds, ants and termites. Affects both the visitors and the durability of the buildings.

The investigation commenced with a visual inspection of the internal and external palace wall. As shown in **Figure 6**, the damage appeared to intensify on the outside wall surface and building components compared with that of the internal wall surface. The visual examination was followed by measuring the slope deflection of the masonry wall as shown in **Figure 7c**.

4. Engineering geology investigation

Gondar city of Ethiopia at an elevation of about 2133 meter above sea level is entirely covered with volcanic rocks and basalts [23]. The engineering geology of the site play an important role in any restoration and conservation activity. Therefore, it becomes very crucial to identify the engineering geology of the palace. So that, accordingly proper remedial measure can be planned. The lack of basic knowledge about geological – geotechnical characteristics of the urban environment and proper planning underlies many several social, geotechnical, and economic issues [24]. Such as, erosion, slope stability, landslide and building cracks, etc., which endanger peoples and infrastructural networks. Three geomorphological processes (weathering, physical impact, faulting and intrusion) were identified in the area, which are responsible for the formation of the present wall slide and crack

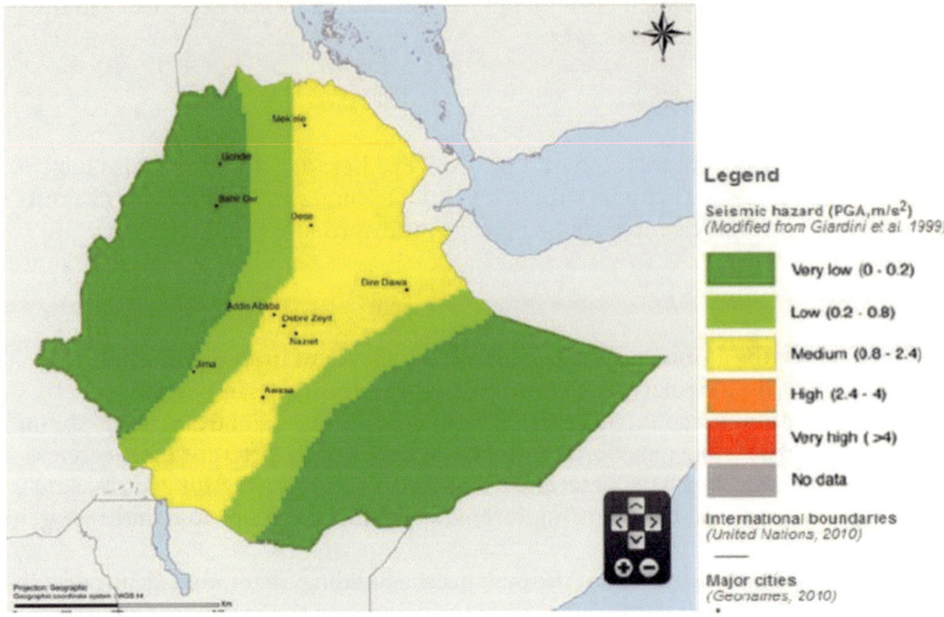

Figure 8.
Seismic hazard classification of Ethiopia [27].

Figure 9.
Iyasu Castle Fasil at Ghebbi [29].

formations. The first two processes are found to be more active and seasonal actions with existing structures and affect the floor, wall and the wooden extended veranda of the palace.

In seismically active area, and the effects of earthquakes on monuments, buildings and assets are critically destructive, given the weak mechanical behavior of masonry when struck by seismic loads [25]. Unfortunately, only when such seismic events occur do institutions and general public become aware of the serious vulnerability of the national architectural heritage.

According Haile [26] East Africa have three main zones of seismic weaknesses in the crustal segments: the East African rift system, the Gulf of Aden, and along the Red Sea. When we examine the location Seismic classification for the location around Gondar its far from the mentioned high risk areas and the Gondar city is classified as in a range from very low to low seismic hazard [27, 28] (**Figure 8**).

Even if the seismic zonation of Fasil Ghebbi is safe but in there have been a major geological incident in the Fasil Ghebbi as reported by Briggs [29] Iyasu Castle was partially damaged by earthquake in 1704 and the ground floor collapsed under the British bombardment in world war II. **Figure 9** illustrates destructive effects the earthquake that occurred on the Fasil Ghebbi at Iyasu Castle in 1704.

5. Hydrogeology

Brimblecombe [30] states water is a critical factor in damage to heritage, it may even be more critical than temperature at same cases. Based on his survey on heritage managers ranked the potential problems faced in a changing English climate as: (1) Rainfall (2) Flooding and soil moisture content (3) Extreme weather (winds and rainfall) (4) Temperature and relative humidity (5) Pests and diseases.

Climate change and world heritage report states intense rainfall and flooding (sea, river) could cause physical changes to porous building materials and finishes due to rising damp; damage due to faulty or inadequate water disposal systems; historic rain water goods not capable of handling heavy rain and often difficult to access, maintain, and adjust erosion of inorganic, organic materials due to flood waters sub soil instability, ground motion and subsidence relative humidity cycles/shock causing splitting, cracking, flaking and dusting of materials and surfaces. Temperature changes diurnal, seasonal, extreme events could results deterioration of facades due to thermal stress freeze–thaw/frost damage, damage inside brick, stone, ceramics that has got wet and frozen within material before drying, biochemical deterioration [31].

Krauss and Fischer [32] studied endangerment of the Palmyra to flooding using DEM and concluded that Temple of Bel lies on a locally higher area the tombs to the left are located mostly on the mid slopes but also remnants can be found in the endangered lower areas which is exposed to flooding. Using simple digital elevation model generated using *Arc GIS* utilizing elevation obtained from Google Earth it can be seen in **Figure 10**, that most of the structures in the Fasil Ghebbi are situated on the upper elevation therefore the risk of flooding during peak rainfall is minimal.

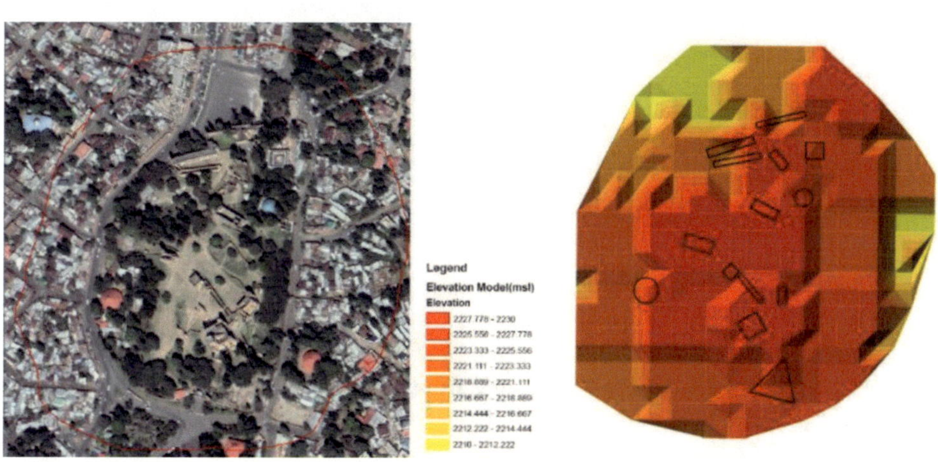

Figure 10.
Fasil Ghebbi digital elevation model.

Figure 11.
Biological growth at Fort San Lorenzo (Panama); source and Fasil Ghebbi [source; Eskinder Desta]. A. Biological growth at Fort San Lorenzo (Panama). B. Biological growth at Fasil Ghebbi.

Figure 12.
Some selected spots inside the royal palace before and after restoration. (A and C): in 2015; (B and D): in 2019 [source; Eskinder Desta].

Historic buildings are more porous and draw water from the ground into their structure and lose it to the environment by surface evaporation. Their wall surfaces and floors are the point of exchange for these reactions. Increases in soil moisture might result in greater salt mobilization and consequent damaging crystallization on decorated surfaces through drying [31].

Climate and biological effects may include changes to lichen colonies on buildings, spread of existing and new species of insects, increase in mound growth. According to UNESCO climate change and world heritage report [31] this factors may cause changes in appearance and collapse of structural timber and timber finishes. As an example; **Figure 11** shows the biological growth at Fort San Lorenzo [33] and Fasil Ghebbi [4] (**Figure 12**).

6. Conclusion

This paper reviews the state-of-the-art knowledge on the durability issue of Fasil Ghebbi palace, Gondar, Ethiopia with long service life. In conclusion it can be said that the protection of the built heritage against chemical changes, prolonged weathering and human intervention threats is a very complex issue, where many factors have to be taken into consideration, among the most important factors are: the traditional construction techniques, the current and future meteorological parameters and the availability of building materials.

The research points out that climate change and human interventions are a real threat to the fail Ghebbi palace. Unless measure is taken for the protection of human impact and weathering related problem on the surface of the building or overall building, these historical heritages will ultimately be endangered. Deterioration related to vegetative organisms was found in Fasil Ghebbi palace; specifically, in rainy seasons these effects were more intensified particularly on the wall that did not face the sun.

Structural instability is also observed for done shaped masonry walls and partitions walls located close to the main traffic road because of the effect of traffic vibration and deliberate mechanical impact. The effect of moisture already has spotted on the appearance of almost all building inside the royal palace, affecting the esthetic value and reducing the bearing capacity of the building.

During the restoration work, it is highly important to select binding materials that exhibit similar properties to the original building material. The research indicates that the Fasil Ghebbi palace is are extremely vulnerable concerning future moisture related durability issue and aging, it is urgently needed to find protection solution and setting a guideline for periodical inspection and restoration work.

Author details

Eskinder Desta Shumuye[1]* and Gashaw Assefa Bezabih[2]

1 School of Civil Engineering, Zhengzhou University, Zhengzhou, China

2 Faculty of Civil Engineering and Built Environment, Construction Technology and Management Department, Hawassa University Institute of Technology, Hawassa, Ethiopia

*Address all correspondence to: eskdes@gs.zzu.edu.cn

References

[1] Aalund F. Master Plan for the Preservation and Presentation of Cultural Heritage. Paris: UNESCO; 1985

[2] Fasil Ghebbi, Gondar Region—UNESCO World Heritage Centre. [Online]. Available from: https://whc.unesco.org/en/list/19 [Accessed: 18 April 2019]

[3] Gebrehiwot M. Ethiopia to renovate four world heritage sites. In: Ethiosports. London, UK: Embassy of Ethiopia; 2017

[4] Shumuye ED. Investigation of the binding materials properties and assessment of durability issue in Fasil 'Ghibbi' palace in Gondar, Ethiopia. Engineering Failure Analysis. **9**(3):10

[5] Wright GRH. Ancient Building Technology. Leiden; Boston: Brill; 2000

[6] Akçaözoğlu K, Fener M, Akçaözoğlu S, Öcal R. Microstructural examination of the effect of elevated temperature on the concrete containing clinoptilolite. Construction and Building Materials. 2014;**72**:316-325

[7] Elena J, Lucia MD. Application of X ray diffraction (XRD) and scanning electron microscopy (SEM) methods to the Portland cement hydration processes. Journal of Applied Engineering Sciences. 2012;**2**(15):35-42

[8] Borges C, Santos Silva A, Veiga R. Durability of ancient lime mortars in humid environment. Construction and Building Materials. 2014;**66**:606-620

[9] Moncmanová A. Environmental factors that influence the deterioration of materials. In: WIT Transactions on State of the Art in Science and Engineering. 1st ed. Vol. 1, A. Southampton, UK: Moncmanová, Ed. WIT Press; 2007. pp. 1-25

[10] Clim D-A, Groll L, Diaconu L-I. Moisture—The main cause of the degradation of historic buildings. Bulletin of the Political Institute In Lasi. 2017;**63**:14

[11] Gupta SP. Climate Change and its Impact on Monumental and Historical Buildings with Reference to Monuments of Chhattisgarh. European Chemical Bulletin. 2013;**2**(8):576-578

[12] Ayehu N. Numerical Groundwater Flow Modeling of the Northern River Catchment of the Lake Tana. Ethiopia: Addis Ababa University; 2010

[13] Gonder Climate: Average Temperature, Weather by Month, Gonder Weather Averages—Climate-Data.org. [Online]. Available from: https://en.climate-data.org/africa/ethiopia/amhara/gonder-1183/ [Accessed: 08 May 2019]

[14] Saba M, Quiñones-Bolaños EE, Martínez Batista HF. Impact of environmental factors on the deterioration of the wall of Cartagena de Indias. Journal of Cultural Heritage. 2019:S129620741830709X

[15] Korkanç M. Deterioration of different stones used in historical buildings within Nigde province, Cappadocia. Construction and Building Materials. 2013;**48**:789-803

[16] Ortega-Calvo JJ, Hernandez-Marine M, Saiz-Jimenez C. Biodeterioration of building materials by cyanobacteria and algae. International Biodeterioration. 1991;**28**(1-4):165-185

[17] van Hees RPJ. 3. Damage analysis as a step towards compatible repair mortars. In: RILEM Report 28: Characterisation of Old Mortars with Respect to their Repair, RILEM Publications SARL. 2004. pp. 107-152

[18] Palomo A, Blanco-Varela MT, Martinez-Ramirez S, Puertas F, Fortes C. Historic mortars: Characterization and durability. New Tendencies for Research; 2014:21

[19] Han S-J, Yoo M, Kim D-W, Wee J-H. Carbon dioxide capture using calcium hydroxide aqueous solution as the absorbent. Energy & Fuels. 2011;**25**(8):3825-3834

[20] Abdul-Rashid R, Ahmad AG. The implementation of maintenance works for historical buildings—A review on the current scenario. Procedia Engineering. 2011;**20**:415-424

[21] Pinho FFS, Lúcio VJG, Faria P, Baião MFC. Durability Aspects Related to Rubble Stone Masonry Walls Strengthened with Reinforced Micro-Concrete Layers. International seminar on seismic risk and rehabilitation of stone masonry housing. 2016. p. 11

[22] Baharuddin MN, Bahardin NF, Rashid RA, Hashim H, Ali IM. Analysis of critical factors and difficulties in maintaining historical building—A current implementation. MATEC Web of Conferences. 2014;**15**:01012

[23] Kassaye S. Improving Ethiopia's Image as a Tourist Destination Master's thesis. University of Applied Sciences; 2013. p. 5-50

[24] Pons NAD, Pejon OJ, Zuquette LV. Use of geoprocessing in the study of land degradation in urban environments: The case of the city of São Carlos, state of São Paulo, Brazil. Environmental Geology. 2007;**53**(4):727-739

[25] Borri A, Corradi M. Architectural heritage: A discussion on conservation and safety. Heritage. 2019;**2**(1):631-647

[26] Haile M. Seismic microzonation for the city of Addis Ababa by using microtremors. presented at the 13th World Conference on Earthquake Engineering; Vancouver; August 1-6, 2004

[27] Herbert S. Assessing Seismic Risk in Ethiopia. (GSDRC Helpdesk Research Report 1087). Governance and social development resource center. Birmingham, UK: University of Birmingham; 2013. p. 8-12

[28] Gouin P. Evaluating Seismic Activity in Ethiopia. (A note for Architects and Engineers). Addis Ababa, Ethiopia: Haile Selassie I University; 1978. p. 5-9

[29] Briggs P, Blatt B. Ethiopia, First. England: Bradt Travel Guides Ltd.; 2007

[30] Brimblecombe P. Refining climate change threats to heritage. Journal of the Institute of Conservation. 2014;**37**(2):85-93

[31] Augustin C. Climate Change and World Heritage. Paris, France: UNESCO World Heritage Centre; 2007. p. 55

[32] Krauss T, Fischer P. Endangerment of cultural heritage sites by heavy rain. In: Fifth International Conference on Remote Sensing and Geoinformation of the Environment (RSCy2017), Paphos, Cyprus. 2017. p. 6

[33] Bonazza A, Palazzi E, Ciantelli C, von Hardenberg J. How can climate change affect the UNESCO cultural heritage sites in Panama? Geoscience. 2018;**8**:7